I0417496

Little Essays - Of Love and Virtue

HAVELOCK ELLIS

Published in 1922

TABLE OF CONTENTS

PREFACE

In these Essays—little, indeed, as I know them to be, compared to the magnitude of their subjects—I have tried to set forth, as clearly as I can, certain fundamental principles, together with their practical application to the life of our time. Some of these principles were stated, more briefly and technically, in my larger Studies of sex; others were therein implied but only to be read between the lines. Here I have expressed them in simple language and with some detail. It is my hope that in this way they may more surely come into the hands of young people, youths and girls at the period of adolescence, who have been present to my thoughts in all the studies I have written of sex because I was myself of that age when I first vaguely planned them. I would prefer to leave to their judgment the question as to whether this book is suitable to be placed in the hands of older people. It might only give them pain. It is in youth that the questions of mature age can alone be settled, if they ever are to be settled, and unless we begin to think about adult problems when we are young all our thinking is likely to be in vain. There are but few people who are able when youth is over either on the one hand to re-mould themselves nearer to those facts of Nature and of Society they failed to perceive, or had not the courage to accept, when they were young, or, on the other hand, to mould the facts of the exterior world nearer to those of their own true interior world. One hesitates to bring home to them too keenly what they have missed in life. Yet, let us remember, even for those who have missed most, there always remains the fortifying and consoling thought that they may at least help to make the world better for those who come after them, and the possibilities of human adjustment easier for others than it has been for themselves.

They must still remain true to their own traditions. We could not wish it to be otherwise.

The art of making love and the art of being virtuous;—two aspects of the great art of living that are, rightly regarded, harmonious and not at variance—remain, indeed, when we cease to misunderstand them, essentially the same in all ages and among all peoples. Yet, always and everywhere, little modifications become necessary, little, yet, like so many little things, immense in their significance and results. In this way, if we are really alive, we flexibly adjust ourselves to the world in which we find ourselves, and in so doing simultaneously adjust to ourselves that ever-changing world, ever-changing, though its changes are within such narrow limits that it yet remains substantially the same. It is with such modification that we are concerned in these Little Essays.

H.E.

London, 1921

CHILDREN AND PARENTS

The twentieth century, as we know, has frequently been called "the century of the child." When, however, we turn to the books of Ellen Key, who has most largely and sympathetically taken this point of view, one asks oneself whether, after all, the child's century has brought much to the child. Ellen Key points out, with truth, that, even in our century, parents may for the most part be divided into two classes: those who act as if their children existed only for their benefit, and those who act as if they existed only for their children's benefit, the results, she adds being alike deplorable. For the first group of parents tyrannise over the child, seek to destroy its individuality, exercise an arbitrary discipline too spasmodic to have any of the good effects of discipline and would model him into a copy of themselves, though really, she adds, it ought to pain them very much to see themselves exactly copied. The second group of parents may wish to model their children not after themselves but after their ideals, yet they differ chiefly from the first class by their over-indulgence, by their anxiety to pamper the child by yielding to all his caprices and artificially protecting him from the natural results of those caprices, so that instead of learning freedom, he has merely acquired self-will. These parents do not indeed tyrannise over their children but they do worse; they train their children to be tyrants. Against these two tendencies of our century Ellen Key declares her own Alpha and Omega of the art of education. Try to leave the child in peace; live your own life beautifully, nobly, temperately, and in so living you will sufficiently teach your children to live.

It is not my purpose here to consider how far this conception of the duty of parents towards children is justified, and whether or not peace is the best preparation for a world in which struggle dominates. All these questions about education are rather idle. There are endless theories of education but no agreement concerning the value of any of them, and the whole question of education remains open. I am here concerned less with the duty of parents in relation to their children than with the duty of children in relation to their parents, and that means that I am not concerned with young children, to whom, that duty still presents no serious problems, since they have not yet developed a personality with self-conscious individual needs. Certainly the one attitude must condition the other attitude. The reaction of children against their parents is the necessary result of the parents' action. So that we have to pay some attention to the character of parental action.

We cannot expect to find any coherent or uniform action on the part of parents. But there have been at different historical periods different general tendencies in the attitude of parents towards their children. Thus if we go back four or five centuries in English social history we seem to find a general attitude which scarcely corresponds exactly to either of Ellen Key's two groups. It seems usually to have been compounded of severity and independence; children were first strictly compelled to go their parents' way and then thrust off to their own way. There seems a certain hardness in this method, yet it is doubtful whether it can fairly be regarded as more unreasonable than either of the two modern methods deplored by Ellen Key. On the contrary it had points for admiration. It was primarily a discipline, but it was regarded, as any fortifying discipline should be regarded, as a preparation for freedom, and it is precisely there that the more timid and clinging modern way seems to fail.

We clearly see the old method at work in the chief source of knowledge concerning old English domestic life, the Paston Letters. Here we find that at an early age the sons of knights and gentlemen were sent to serve in the houses of other gentlemen: it was here that their education really took place, an education not in book knowledge, but in knowledge of life. Such education was considered so necessary for a youth that a father who kept his sons at home was regarded as negligent of his duty to his family. A knowledge of the world was a necessary part, indeed the chief part, of a youth's training for life. The remarkable thing is that this applied also to a large extent to the daughters. They realised in those days, what is only beginning to be realised in ours,[1] that, after all, women live in the world just as much, though differently, as men live in the world, and that it is quite as necessary for the girl as for the boy to be trained to the meaning of life. Margaret Paston, towards the end of the fifteenth century, sent her daughter Ann to live in the house of a gentleman who, a little later, found that he could not keep her as he was purposing to decrease the size of his household. The mother writes to her son: "I shall be fain to send for her and with me she shall but lose her time, and without she be the better occupied she shall oftentimes move me and put me to great unquietness. Remember what labour I had with your sister, therefore do your best to help her forth"; as a result it was planned to send her to a relative's house in

London.

[1] This was illustrated in England when women first began to serve on juries. The pretext was frequently brought forward that there are certain kinds of cases and of evidence that do not concern women or that women ought not to hear. The pretext would have been more plausible if it had also been argued that there are certain kinds of cases and of evidence that men ought not to hear. As a matter of fact, whatever frontier there may be in these matters is not of a sexual kind. Everything that concerns men ultimately concerns women, and everything that concerns women ultimately concerns men. Neither women nor men are entitled to claim dispensation.

It is evident that in the fifteenth century in England there was a wide prevalence of this method of education, which in France, a century later, was still regarded as desirable by Montaigne. His reason for it is worth noting; children should be educated away from home, he remarks, in order to acquire hardness, for the parents will be too tender to them. "It is an opinion accepted by all that it is not right to bring up children in their parents' laps, for natural love softens and relaxes even the wisest."[2]

[2] Montaigne, Essais, Bk. I., ch. 25.

In old France indeed the conditions seem similar to those in England. The great serio-comic novel of Antoine de la Salle, Petit Jean de Saintré, shows us in detail the education and the adventures, which certainly involved a very early introduction to life, of a page in a great house in the fifteenth century. We must not take everything in this fine comedy too solemnly, but in the fourteenth century Book of the Knight of the Tour-Landry we may be sure that we have at its best the then prevailing view of the relation of a father to his tenderly loved daughters. Of harshness and rigour in the relationship it is not easy to find traces in this lengthy and elaborate book of paternal counsels. But it is clear that the father takes seriously the right of a daughter to govern herself and to decide for herself between right and wrong. It is his object, he tells his girls, "to enable them to govern themselves." In this task he assumes that they are entitled to full knowledge, and we feel that he is not instructing them in the mysteries of that knowledge; he is taking for granted, in the advice he gives and the stories he tells them, that his "young and small daughters, not, poor things, overburdened with experience," already possess the most precise knowledge of the intimate facts of life, and that he may tell them, without turning a hair, the most outrageous incidents of debauchery. Life already lies naked before them: that he assumes; he is not imparting knowledge, he is giving good counsel.[3]

[3] If the Knight went to an extreme in his assumption of his daughters' knowledge, modern fathers often go to the opposite and more foolish extreme of assuming in their daughters an ignorance that would be dangerous even if it really existed. In A Young Girl's Diary (translated from

the German by Eden and Cedar Paul), a work that is highly instructive for parents, and ought to be painful for many, we find the diarist noting at the age of thirteen that she and a girl friend of about the same age overheard the father of one of them—both well brought up and carefully protected, one Catholic and the other Protestant—referring to "those innocent children." "We did laugh so, WE and innocent children!!! What our fathers really think of us; we innocent!!! At dinner we did not dare look at one another or we should have exploded." It need scarcely be added that, at the same time, they were more innocent than they knew.

It is clear that this kind of education and this attitude towards children must be regarded as the outcome of the whole mediæval method of life. In a state of society where roughness and violence, though not, as we sometimes assume, chronic, were yet always liable to be manifested, it was necessary for every man and woman to be able to face the crudest facts of the world and to be able to maintain his or her own rights against them. The education that best secured that strength and independence was the best education and it necessarily involved an element of hardness. We must go back earlier than Montaigne's day, when the conditions were becoming mitigated, to see the system working in all its vigour.

The lady of the day of the early thirteenth century has been well described by Luchaire in his scholarly study of French Society in the time of Philip Augustus. She was, he tells us, as indeed she had been in the preceding feudal centuries, often what we should nowadays call a virago, of violent temperament, with vivid passions, broken in from childhood to all physical exercises, sharing the pleasures and dangers of the knights around her. Feudal life, fertile in surprises and in risks, demanded even in women a vigorous temper of soul and body, a masculine air, and habits also that were almost virile. She accompanied her father or her husband to the chase, while in war-time, if she became a widow or if her husband was away at the Crusades, she was ready, if necessary, to direct the defences of the lordship, and in peace time she was not afraid of the longest and most dangerous pilgrimages. She might even go to the Crusades on her own account, and, if circumstances required, conduct a war to come out victoriously.

We may imagine the robust kind of education required to produce people of this quality. But as regards the precise way in which parents conducted that education, we have, as Luchaire admits, little precise knowledge. It is for the most part only indirectly, by reading between the lines, that we glean something as to what it was considered befitting to inculcate in a good household, and as what we thus learn is mostly from the writings of Churchmen it is doubtless a little one-sided. Thus Adam de Perseigne, an ecclesiastic, writes to the Countess du Perche to advise her how to live in a Christian manner; he counsels her to abstain from playing games of chance and chess, not to take pleasure in the indecent farces of actors, and to be moderate in dress. Then, as ever, preachers expressed their horror of the ruinous extravagance of women, their false hair, their rouge, and their dresses that were too long or too short. They also reprobated their love of flirtation. It was, however, in those days a young girl's recognised duty, when a knight arrived in the household, to exercise the rites of hospitality, to disarm him, give him his bath, and if necessary massage him to help him to go to sleep. It is not surprising that the young girl sometimes made love to the knight under these circumstances, nor is it surprising that he, engaged in an arduous life and trained to disdain feminine attractions, often failed to respond.

It is easy to understand how this state of things gradually became transformed into the considerably different position of parents and child we have known, which doubtless attained its climax nearly a century ago. Feudal conditions, with the large households so well adapted to act as seminaries for youth, began to decay, and as education in such seminaries must have led to frequent mischances both for youths and maidens who enjoyed the opportunities of education there, the regret for their disappearance may often have been tempered for parents. Schools, colleges, and universities began to spring up and develop for one sex, while for the other home life grew more intimate, and domestic ties closer. Montaigne's warning against the undue tenderness of a narrow family life no longer seemed reasonable, and the family became more self-centred and more enclosed. Beneath this, and more profoundly influential, there was a general softening in social respects, and a greater expansiveness of affectional relationships, in reality or in seeming, within the home, compensating, it may be, the more diffused social feeling within a group which characterised the previous period.

So was cultivated that undue tenderness, deplored by Montaigne, which we now regard as almost normal in family life, and solemnly label, if we happen to be psycho-analysts, the Oedipus-complex or the Electra-complex. Sexual love is closely related to parental love; the tender emotion, which is an intimate part of parental love, is also an intimate part of sexual love, and two emotions which are each closely related to a third emotion cannot fail to become often closely associated to each other. With a little thought we might guess beforehand, even while still in complete ignorance of the matter, that there could not fail to be frequently a sexual tinge in the affection of a father for his daughter, of a mother for her son, of a son for his mother, or a daughter for her father. Needless to say, that does not mean that there is present any physical desire of sex in the narrow sense; that would be a perversity, and a rare perversity. We are here on another plane than that of crude physical desire, and are moving within the sphere of the emotions. But such emotions are often strong, and all the stronger because conscious of their own absolute rectitude and often masked under the shape of Duty. Yet when prolonged beyond the age of childhood they tend to become a clog on development, and a hindrance to a wholesome life. The child who cherishes such emotion is likely to suffer infantile arrest of development, and the parent who is so selfish as to continue to expend such tenderness on a child who has passed the age of childhood, or to demand it, is guilty of a serious offence against that child.

That the intimate family life which sometimes resulted—especially when, as frequently happened, the seeming mutual devotion was also real—might often be regarded as beautiful and almost ideal, it has been customary to repeat with an emphasis that in the end has even become nauseous. For it was usually overlooked that the self-centred and enclosed family, even when the mutual affection of its members was real enough to bear all examination, could scarcely be more than partially beautiful, and could never be ideal. For the family only represents one aspect, however important an aspect, of a human being's functions and activities. He cannot, she cannot, be divorced from the life of the social group, and a life is beautiful and ideal, or the reverse, only when we have taken into our consideration the social as well as the family relationship. When the family claims to prevent the free association of an adult member of it with the larger social organisation, it is claiming that the part is greater than the whole, and such a claim cannot fail to be morbid and mischievous.

The old-world method of treating children, we know, has long ago been displaced as containing an element of harsh tyranny. But it was not perceived, and it seems indeed not even yet to be generally recognised, that the system which replaced it, and is only now beginning to pass away, involved another and more subtle tyranny, the more potent because not seemingly harsh. Parents no longer whipped their children even when grown up, or put them in seclusion, or exercised physical force upon them after they had passed childhood. They felt that that would not be in harmony with the social customs of a world in which ancient feudal notions were dead. But they merely replaced the external compulsion by an internal compulsion which was much more effective. It was based on the moral assumption of claims and duties which were rarely formulated because parents found it quite easy and pleasant to avoid formulating them, and children, on the rare occasions when they formulated them, usually felt a sense of guilt in challenging their validity. It was in the nineteenth century that this state of things reached its full development. The sons of the family were usually able, as they grew up, to escape and elude it, although they thereby often created an undesirable divorce from the home, and often suffered, as well as inflicted, much pain in tearing themselves loose from the spiritual bonds—especially perhaps in matters of religion—woven by long tradition to bind them to their parents. It was on the daughters that the chief stress fell. For the working class, indeed, there was often the possibility of escape into hard labour, if only that of marriage. But such escape was not possible, immediately or at all, for a large number. During the nineteenth century many had been so carefully enclosed in invisible cages, they had been so well drilled in the reticences and the duties and the subserviences that their parents silently demanded of them, that we can never know all the tragedies that took place. In exceptional cases, indeed, they gave a sign. When they possessed unusual power of intellect, or unusual power of character and will, they succeeded in breaking loose from their cages, or at least in giving expression to themselves. This is seen in the stories of nearly all the women eminent in life and literature during the nineteenth century, from the days of Mary Wollstonecraft onwards. The Brontës, almost, yet not quite, strangled by the fetters placed upon them by their stern and narrow-minded father, and enabled to attain the full stature of their genius only by that brief sojourn in Brussels, are representative. Elizabeth Barrett, chained to a couch of invalidism under the eyes of an imperiously affectionate father until with Robert Browning's aid she secretly eloped into the open air of freedom and health, and so attained complete literary expression, is a typical figure. It is only because we recognise that she is a typical figure among the women who attained distinction that we are able to guess at the vast number of mute inglorious Elizabeth Barretts who were never able to escape by their own efforts and never found a Browning to aid them to escape.

It is sometimes said that those days are long past and that young women, in all the countries which we are pleased to called civilised, are now emancipated, indeed, rather too much emancipated. Critics come forward to complain of their undue freedom, of their irreverent familiarity to their parents, of their language, of their habits. But there were critics who said the very same things, in almost the same words, of the grandmothers of these girls! These incompetent critics are as ignorant of the social history of the past as they are of the social significance of the history of the present. We read in Once a Week of sixty years ago (10th August, 1861), the very period when the domestic conditions of girls were the most oppressive in the sense here understood, that these same critics were about at that time, and as shocked as they are now at "the young ladies who talk of 'awful swells' and 'deuced bores,' who smoke and venture upon free discourse, and try to be like men." The writer of this anonymous article, who was really (I judge from internal evidence) so distinguished and so serious a woman as Harriet Martineau, duly snubs these critics, pointing out that such accusations are at least as old as Addison and Horace Walpole; she remarks that there have no doubt been so-called "fast young ladies" in every age, "varying their doings and sayings according to the fopperies of the time." The question, as she pertinently concludes is, as indeed it still remains to-day: "Have we more than the average proportion? I do not know." Nor to-day do we know.

But while to-day, as ever before, we have a certain proportion of these emancipated girls, and while to-day, as perhaps never before, we are able to understand that they have an element of reason on their side, it would be a mistake to suppose that they are more than exceptions. The majority are unable, and not even anxious, to attain this light-hearted social emancipation. For the majority, even though they are workers, the anciently subtle ties of the home are still, as they should be, an element of natural piety, and, also, as they should not be, clinging fetters which impede individuality and destroy personal initiative.

We all know so many happy homes beneath whose calm surface this process is working out. The parents are deeply attached to their children, who still remain children to them even when they are grown up. They wish to guide them and mould them and cherish them, to protect them from the world, to enjoy their society and their aid, and they expect that their children shall continue indefinitely to remain children. The children, on their side, remain and always will remain, tenderly attached to their parents, and it would really pain them to feel that they are harbouring any unwillingness to stay in the home even after they have grown up, so long as their parents need their attention. It is, of course, the daughters who are thus expected to remain in the home and who feel this compunction about leaving it. It seems to us—although, as we have seen, so unlike the attitude of former days—a natural, beautiful, and rightful feeling on both sides.

Yet, in the result, all sorts of evils tend to ensue. The parents often take as their moral right the services which should only be accepted, if accepted at all, as the offering of love and gratitude, and even reach a degree of domineering selfishness in which they refuse to believe that their children have any adult rights of their own, absorbing and drying up that physical and spiritual life-blood of their offspring which it is the parents' part in Nature to feed. If the children are willing there is nothing to mitigate this process; if they are unwilling the result is often a disastrous conflict. Their time and energy are not their own; their tastes are criticised and so far as possible crushed; their political ideas, if they have any, are treated as pernicious; and—which is often on both sides the most painful of all— differences in religious belief lead to bitter controversy and humiliating recrimination. Such differences in outlook between youth and age are natural and inevitable and right. The parents themselves, though they may have forgotten it, often in youth similarly revolted against the cherished doctrines of their own parents; it has ever been so, the only difference being that to-day, probably, the opportunities for variation are greater. So it comes about that what James Hinton said half a century ago is often true to-day: "Our happy Christian homes are the real dark places of the earth."

It is evident that the problem of the relation of the child to the parent is still incompletely solved even in what we consider our highest civilisation. There is here needed an art in which those who have to exercise it can scarcely possess all the necessary skill and experience. Among trees and birds and beasts the art is surer because it is exercised unconsciously, on the foundation of a large tradition in which failure meant death. In the common procreative profusion of those forms of life the frequent death of the young was a matter of little concern, but biologically there was never any sacrifice of the offspring to the well-being of the parents. Whenever sacrifice is called for it is the parents who are sacrificed to their offspring. In our superior human civilisation, in which quantity ever tends to give place to quality, the higher value of the individual involves an effort to avoid sacrifice which sometimes proves worse than abortive. An avian philosopher would be unlikely to feel called upon to denounce nests as the dark places of the earth, and in laying down our human moral laws we have always to be aware of forgetting the fundamental biological relationship of parent and child to which all such moral laws must conform. To some would-be parents that necessity may seem hard. In such a case it is well for them to remember that there is no need to become parents and that we live in an age when it is not difficult to avoid becoming a parent. The world is not dying for lack of parents. On the contrary we have far too many of them—ignorant parents, silly parents, unwilling parents, undesirable parents—and those who aspire to the high dignity of creating the future race, let them be as few as they will—and perhaps at the present time the fewer the better—must not refuse the responsibilities of that position, its pains as well as its joys.

In our human world, as we know, the moral duties laid upon us—the duties in which, if we fail, we become outcasts in our own eyes or in those of others or in both—are of three kinds: the duties to oneself, the duties to the small circle of those we love, and the duties to the larger circle of mankind to which ultimately we belong, since out of it we proceed, and to it we owe all that we are. There are no maxims, there is only an art and a difficult art, to harmonise duties which must often conflict. We have to be true to all the motives that sanctify our lives. To that extent George Eliot's Maggie Tulliver was undoubtedly right. But the renunciation of the Self is not the routine solution of every conflict, any more than is the absolute failure to renounce. In a certain sense the duty towards the self comes before all others, because it is the condition on which duties towards others possess any significance and worth. In that sense, it is true according to the familiar saying of Shakespeare,—though it was only Polonius, the man of maxims, who voiced it,—that one cannot be true to others unless one is first true to oneself, and that one can know nothing of giving aught that is worthy to give unless one also knows how to take.

We see that the problem of the place of parents in life, after their function of parenthood has been adequately fulfilled, a problem which offers no difficulties among most forms of life, has been found hard to solve by Man. At some places and periods it has been considered most merciful to put them, to death; at others they have been almost or quite deified and allowed to regulate the whole lives of their descendants. Thus in New Caledonia aged parents, it is said by Mrs. Hadfield, were formerly taken up to a high mountain and left with enough food to last a few days; there was at the same time great regard for the aged, as also among the Hottentots who asked: "Can you see a parent or a relative shaking and freezing under a cold, dreary, heavy, useless old age, and not think, in pity of them, of putting an end to their misery?" It was generally the opinion of the parents themselves, but in some countries the parents have dominated and overawed their children to the time of their natural death and even beyond, up to the point of ancestor worship, as in China, where no man of any age can act for himself in the chief matters of life during his parents' life-time, and to some extent in ancient Rome, whence an influence in this direction which still exists in the laws and customs of France.[4] Both extremes have proved compatible with a beautifully human life. To steer midway between them seems to-day, however, the wisest course. There ought to be no reason, and under happy conditions there is no reason, why the relationship between parent and child, as one of mutual affection and care, should ever cease to exist. But that the relationship should continue to exist as a tie is unnatural and tends to be harmful. At a certain stage in the development of the child the physical tie with the parent is severed, and the umbilical cord cut. At a later stage in development, when puberty is attained and adolescence is feeling its way towards a complete adult maturity, the spiritual tie must be severed. It is absolutely essential that the young spirit should begin to essay its own wings. If its energy is not equal to this adventure, then it is the part of a truly loving parent to push it over the edge of the nest. Of course there are dangers and risks. But the worst dangers and risks come of the failure to adventure, of the refusal to face the tasks of the world and to assume the full function of life. All that Freud has told of the paralysing and maiming influence of infantile arrest or regression is here profitable to consider. In order, moreover, that the relationship between parents and children may retain its early beauty and love, it is essential that it shall adapt itself to adult conditions and the absence of ties so rendered necessary. Otherwise there is little likelihood of anything but friction and pain on one side or the other, and perhaps on both sides.

[4] The varying customs of different peoples in this matter are set forth by Westermarck, The Origin and Development of the Moral Ideas, Ch. XXV.

The parents have not only to train their children: it is of at least equal importance that they should train themselves. It is desirable that children, as they grow up, should be alive to this necessity, and consciously assist in the process, since they are in closer touch with a new world of activities to which their more lethargic parents are often blind and deaf. For every fresh stage in our lives we need a fresh education, and there is no stage for which so little educational preparation is made as that which follows the reproductive period. Yet at no time—especially in women, who present all the various stages of the sexual life in so emphatic a form—would education be more valuable. The great burden of reproduction, with all its absorbing responsibilities, has suddenly been lifted; at the same time the perpetually recurring rhythm of physical sex manifestations, so often disturbing in its effect, finally ceases; with that cessation, very often, after a brief period of perturbation, there is an increase both in physical and mental energy. Yet, too often, all that one can see is that a vacuum has been created, and that there is nothing to fill it. The result is that the mother—for it is most often of the mother that complaint is made—devotes her own new found energies to the never-ending task of hampering and crushing her children's developing energies. How many mothers there are who bring to our minds that ancient and almost inspired statement concerning those for whom "Satan finds some mischief still"! They are wasting, worse than wasting, energies that might be profitably applied to all sorts of social service in the world. There is nothing that is so much needed as the "maternal in politics," or in all sorts of non-political channels of social service, and none can be better fitted for such service than those who have had an actual experience of motherhood and acquired the varied knowledge that such experience should give. There are numberless other ways, besides social service, in which mothers who have passed the age of forty, providing they possess the necessary aptitudes, can more profitably apply themselves than in hampering, or pampering, their adult children. It is by wisely cultivating their activities in a larger sphere that women whose chief duties in the narrower domestic sphere are over may better ensure their own happiness and the welfare of others than either by fretting and obstructing, or by worrying over, their own children who are no longer children. It is quite true that the children may go astray even when they have ceased to be children. But the time to implant the seeds of virtue, the time to convey a knowledge of life, was when they were small. If it was done well, it only remains to exercise faith and trust. If it was done ill, nothing done later will compensate, for it is merely foolish for a mother who could not educate her children when they were small to imagine that she is able to educate them when they are big.

So it is that the problem of the attitude of the child to its parents circles round again to that of the parents to the child. The wise parent realises that childhood is simply a preparation for the free activities of later life, that the parents exist in order to equip children for life and not to shelter and protect them from the world into which they must be cast. Education, whatever else it should or should not be, must be an inoculation against the poisons of life and an adequate equipment in knowledge and skill for meeting the chances of life. Beyond that, and no doubt in the largest part, it is a natural growth and takes place of itself.

THE MEANING OF PURITY

We live in a world in which, as we nowadays begin to realise, we find two antagonistic streams of traditional platitude concerning the question of sexual purity, both flowing from the far past.

The people who embody one of these streams of tradition, basing themselves on old-fashioned physiology, assume, though they may not always assert, that the sexual products are excretions, to be dealt with summarily like other excretions. That is an ancient view and it was accepted by such wise philosophers of old times as Montaigne and Sir Thomas More. It had, moreover, the hearty support of so eminent a theological authority as Luther, who on this ground preached early marriage to men and women alike. It is still a popular view, sometimes expressed in the crudest terms, and often by people who, not following Luther's example, use it to defend prostitution, though they generally exclude women from its operation, as a sex to whom it fails to apply and by whom it is not required.

But on the other hand we have another stream of platitude. On this side there is usually little attempt either to deny or to affirm the theory of the opposing party, though they would contradict its conclusions. Their theory, if they have one, would usually seem to be that sexual activity is a response to stimulation from without or from within, so that if there is no stimulation there will be no sexual manifestation. They would preach, they tell us, a strenuous ideal; they would set up a wholesome dictate of hygiene. The formula put forward on this basis usually runs: Continence is not only harmless but beneficial. It is a formula which, in one form or another, has received apparently enthusiastic approval in many quarters, even from distinguished physicians. We need not be surprised. A proposition so large and general is not easy to deny, and is still more difficult to reverse; therefore it proves welcome to the people—especially the people occupying public and professional positions—who wish to find the path of least resistance, under pressure of a vigorous section of public opinion. Yet in its vagueness the proposition is a little disingenuous; it condescends to no definitions and no qualifications; it fails even to make clear how it is to be reconciled with any enthusiastic approval of marriage, for if continence is beautiful how can marriage make it cease to be so?

Both these streams of feeling, it may be noted, sprang from a common source far back in the primitive human world. All the emanations of the human body, all the spontaneous manifestations of its activities, were mysterious and ominous to early man, pregnant with terror unless met with immense precautions and surrounded by careful ritual. The manifestations of sex were the least intelligible and the most spontaneous. Therefore the things of sex were those that most lent themselves to feelings of horror and awe, of impurity and of purity. They seemed so highly charged with magic potency that there were no things that men more sought to avoid, yet none to which they were impelled to give more thought. The manifold echoes of that primitive conception of sex, and all the violent reactions that were thus evolved and eventually bound up with the original impulse, compose the streams of tradition that feed our modern world in this matter and determine the ideas of purity that surround us.

At the present day the crude theory of the sexual impulse held on one side, and the ignorant rejection of theory altogether on the other side, are beginning to be seen as both alike unjustified. We begin to find the grounds for a sounder theory. Not indeed that the problems of sex, which go so deeply into the whole personal and social life, can ever be settled exclusively upon physiological grounds. But we have done much to prepare even the loftiest Building of Love when we have attained a clear view of its biological basis.

The progress of chemico-physiological research during recent years has now brought us to new ground for our building. Indeed the image might well be changed altogether, and it might be said that science has entirely transferred the drama of reproduction to a new stage with new actors. Therewith the immense emphasis placed on excretion, and the inevitable reaction that emphasis aroused, both alike disappear. The sexual protagonists are no longer at the surface but within the most secret recesses of the organism, and they appear to science under the name of Hormones or Internal Secretions, always at work within and never themselves condescending to appear at all. Those products of the sexual glands which in both sexes are cast out of the body, and at an immature stage of knowledge appeared to be excretions, are of primary reproductive importance, but, as regards the sexual constitution of the individual, they are of far less importance than the internal secretions of these very same glands. It is, however, by no means only the specifically sexual glands which thus exert a sexual influence within the organism. Other glands in the brain, the throat, and the abdomen,—such as the thyroid and the adrenals,—are also elaborating fermentative secretions to throw into the system. Their mutual play is so elaborate that it is only beginning to be understood. Some internal secretions stimulate, others inhibit, and the same secretions may under different conditions do either. This fact is the source of many degrees and varieties of energy and formative power in the organism. Taken altogether, the internal secretions are the forces which build up the man's and woman's distinctively sexual constitution: the special disposition and growth of hair, the relative development of breasts and pelvis, the characteristic differences in motor activity, the varying emotional desires and needs. It is in the complex play of these secretions that we now seek the explanation of all the peculiarities of sexual constitution, imperfect or one-sided physical and psychic development, the various approximations of the male to female bodily and emotional disposition, of the female to the male, all the numerous gradations that occur, naturally as we now see, between the complete man and the complete woman.

When we turn the light of this new conception on to our old ideas of purity,—to the virtue or the vice, accordingly as we may have been pleased to consider it, of sexual abstinence,—we begin to see that those ideas need radical revision. They appear in a new light, their whole meaning is changed. No doubt it may be said they never had the validity they appeared to possess, even when we judge them by the crudest criterion, that of practice. Thus, while it is the rule for physicians to proclaim the advantages of sexual continence, there is no good reason to believe that they have themselves practised it in any eminent degree. A few years ago an inquiry among thirty-five distinguished physicians, chiefly German and Russian, showed that they were nearly all of opinion that continence is harmless, if not beneficial. But Meirowsky found by inquiry of eighty-six physicians, of much the same nationalities, that only one had himself been sexually abstinent before marriage. There seem to be no similar statistics for the English-speaking countries, where there exists a greater modesty—though not perhaps notably less need for it—in the making of such confessions. But if we turn to the allied profession which is strongly on the side of sexual abstinence, we find that among theological students, as has been shown in the United States, while prostitution may be infrequent, no temptation is so frequent or so potent, and in most cases so irresistible, as that to solitary sexual indulgence. Such is the actual attitude towards the two least ideal forms of sexual practice—as distinguished from mere theory—on the part of the two professions which most definitely pronounce in favour of continence.

It is necessary, however, as will now be clearer, to set our net more widely. We must take into consideration every form and degree of sexual manifestation, normal and abnormal, gross and ethereal. When we do this, even cautiously and without going far afield, sexual abstinence is found to be singularly elusive. Rohleder, a careful and conscientious investigator, has asserted that such abstinence, in the true and complete sense, is absolutely non-existent, the genuine cases in which sexual phenomena of some kind or other fail to manifest themselves being simply cases of inborn lack of sexual sensibility. He met, indeed, a few people who seemed exceptions to the general rule, but, on better knowledge, he found that he was mistaken, and that so far from being absent in these people the sexual instinct was present even in its crudest shapes. The activity of sex is an activity that on the physical side is generated by the complex mechanism of the ductless glands and displayed in the whole organism, physical and psychic, of the individual, who cannot abolish that activity, although to some extent able to regulate the forms in which it is manifested, so that purity cannot be the abolition or even the indefinite suspension of sexual manifestations; it must be the wise and beautiful control of them.

It is becoming clear that the old platitudes can no longer be maintained, and that if we wish to improve our morals we must first improve our knowledge.

We have seen that various popular beliefs and conventional assumptions concerning the sexual impulse can no longer be maintained. The sexual activities of the organism are not mere responses to stimulation, absent if we choose to apply no stimulus, never troubling us if we run away from them, harmless if we enclose them within a high wall. Nor do they constitute a mere excretion, or a mere appetite, which we can control by a crude system of hygiene and dietetics. We better understand the psycho-sexual constitution if we regard the motive power behind it as a dynamic energy, produced and maintained by a complex mechanism at certain inner foci of the body, and realise that whatever periodic explosive manifestations may take place at the surface, the primary motive source lies in the intimate recesses of the organism, while the outcome is the whole physical and spiritual energy of our being under those aspects which are most forcible and most aspiring and even most ethereal.

This conception, we find, is now receiving an admirable and beautifully adequate physical basis in the researches of distinguished physiologists in various lands concerning the parts played by the ductless glands of the body, in sensitive equilibrium with each other, pouring out into the system stimulating and inhibiting hormones, which not only confer on the man's or woman's body those specific sexual characters which we admire but at the same time impart the special tone and fibre and polarity of masculinity or femininity to the psychic disposition. Yet, even before Brown-Séquard's first epoch-making suggestion had set physiologists to search for internal secretions, the insight of certain physicians on the medico-psychological side was independently leading towards the same dynamic conception. In the middle of the last century Anstie, an acute London physician, more or less vaguely realised the transformations of sexual energy into nervous disease and into artistic energy. James Hinton, whose genius rendered him the precursor of many modern ideas, had definitely grasped the dynamic nature of sexual activity, and daringly proposed to utilise it, not only as a solution of the difficulties of the personal life but for the revolutionary transformation of morality.[5] It was the wish to group together all the far-flung manifestations of the inner irresistible process of sexual activity that underlay my own conception of auto-erotism, or the spontaneous erotic impulse which arises from the organism apart from all definite external stimulation, to be manifested, or it may be transformed, in mere solitary physical sex activity, in dreams of the night, in day-dreams, in shapes of literature and art, in symptoms of nervous disorder such as some forms of hysteria, and even in the most exalted phases of mystical devotion. Since then, a more elaborate attempt to develop a similar dynamic conception of sexual activity has been made by Freud; and the psycho-analysts who have followed him, or sometimes diverged, have with endless subtlety, and courageous thoroughness, traced the long and sinuous paths of sexual energy in personality and in life, indeed in all the main manifestations of human activity.

[5] "The man who separated the thought of chastity from Service and made it revolve round Self," wrote Hinton half a century ago in his unpublished MSS., "betrayed the human race." "The rule of Self," he wrote again, "has two forms: Self-indulgence and Self-virtue; and Nature has two weapons against it: pain and pleasure.... A restraint must always be put away when another's need can be served by putting it away; for so is restored to us the force by which Life is made.... How curious it seems! the true evil things are our good things. Our thoughts of duty and goodness and chastity, those are the things that need to be altered and put aside; these are the barriers to true goodness.... I foresee the positive denial of all positive morals, the removal of all restrictions. I feel I do not know what 'license,' as we should term it, may not truly belong to the perfect state of Man. When there is no self surely there is no restriction; as we see there is none in Nature.... May we not say of marriage as St. Augustine said of God: 'Rather would I, not finding, find Thee, than finding, not find Thee'?... 'Because we like' is the sole legitimate and perfect motive of human action.... If this is what Nature affirms then it will be what I believe." This dynamic conception of the sexual impulse, as a force that, under natural conditions, may be trusted to build up a new morality, obviously belongs to an indefinitely remote future. It is a force whose blade is two-edged, for while it strikes at unselfishness it also strikes at selfishness, and at present we cannot easily conceive a time when "there is no self"; we should be more disposed to regard it as a time when there is much humbug. Yet for the individual this conception of the

constructive power of love retains much enlightenment and inspiration.

It is important for us to note about this dynamic sexual energy in the constitution that while it is very firmly and organically rooted, and quite indestructible, it assumes very various shapes. On the physical side all the characters of sexual distinction and all the beauties of sexual adornment are wrought by the power furnished by the co-operating furnaces of the glands, and so also, on the psychic side, are emotions and impulses which range from the simplest longings for sensual contact to the most exalted rapture of union with the Infinite. Moreover, there is a certain degree of correlation between the physical and the psychic manifestation of sexual energy, and, to some extent, transformation is possible in the embodiment of that energy.

A vague belief in the transformation of sexual energy has long been widespread. It is apparently shown in the idea that continence, as an economy in the expenditure of sexual force, may be practised to aid the physical and mental development, while folklore reveals various sayings in regard to the supposed influence of sexual abstinence in the causation of insanity. There is a certain underlying basis of reason in such beliefs, though in an unqualified form they cannot be accepted, for they take no account of the complexity of the factors involved, of the difficulty and often impossibility of effecting any complete transformation, either in a desirable or undesirable direction, and of the serious conflict which the process often involves. The psycho-analysts have helped us here. Whether or not we accept their elaborate and often shifting conceptions, they have emphasised and developed a psychological conception of sexual energy and its transformations, before only vaguely apprehended, which is now seen to harmonise with the modern physiological view.

The old notion that sexual activity is merely a matter of the voluntary exercise, or abstinence from exercise, of the reproductive functions of adult persons has too long obstructed any clear vision of the fact that sexuality, in the wide and deep sense, is independent of the developments of puberty. This has long been accepted as an occasional and therefore abnormal fact, but we have to recognise that it is true, almost or quite normally, even of early childhood. No doubt we must here extend the word "sexuality"[6]—in what may well be considered an illegitimate way—to cover manifestations which in the usual sense are not sexual or are at most called "sexual perversions." But this extension has a certain justification in view of the fact that these manifestations can be seen to be definitely related to the ordinary adult forms of sexuality. However we define it, we have to recognise that the child takes the same kind of pleasure in those functions which are natural to his age as the adult is capable of taking in localised sexual functions, that he may weave ideas around such functions, sometimes cultivate their exercise from love of luxury, make them the basis of day-dreams which at puberty, when the ideals of adult life are ready to capture his sexual energy, he begins to grow ashamed of.

[6] Perhaps, as applied to the period below puberty, it would be more exact to say "pseudo-sexuality." Matsumato has lately pointed out the significance of the fact that the interstitial testicular tissue, essential to the hormonic function of the testes, only becomes active at puberty.

At this stage, indeed, we reach a crucial point, though it has usually been overlooked, in the lives of boys and girls, more especially those whose heredity may have been a little tainted or their upbringing a little twisted. For it is here that the transformation of energy and the resulting possibilities of conflict are wont to enter. In the harmoniously developing organism, one may say, there is at this period a gradual and easy transmutation of the childish pleasurable activities into adult activities, accompanied perhaps by a feeling of shame for the earlier feelings, though this quickly passes into a forgetfulness which often leads the adult far astray when he attempts to understand the psychic life of the child. The childish manifestations, it must be remarked, are not necessarily unwholesome; they probably perform a valuable function and develop budding sexual emotions, just as the petals of flowers are developed in pale and contorted shapes beneath the enveloping sheaths.

But in our human life the transmutation is often not so easy as in flowers. Normally, indeed, the adolescent transformations of sex are so urgent and so manifold—now definite sensual desire, now muscular impulses of adventure, now emotional aspirations in the sphere of art or religion—that they easily overwhelm and absorb all its vaguer and more twisted manifestations in childhood. Yet it may happen that by some aberration of internal development or of external influence this conversion of energy may at one point or another fail to be completely effected. Then some fragment of infantile sexuality survives, in rare cases to turn all the adult faculties to its service and become reckless and triumphant, in minor and more frequent cases to be subordinated and more or less repressed into the subconscious sphere by voluntary or even involuntary and unconscious effort. Then we may have conflict, which, when it works happily, exerts a fortifying and ennobling influence on character, when more unhappily a disturbing influence which may even lead to conditions of definite nervous disorder.

The process by which this fundamental sexual energy is elevated from elementary and primitive forms into complex and developed forms is termed sublimation, a term, originally used for the process of raising by heat a solid substance to the state of vapour, which was applied even by such early writers as Drayton and Davies in a metaphorical and spiritual sense.[7] In the sexual sphere sublimation is of vital importance because it comes into question throughout the whole of life, and our relation to it must intimately affect our conception of morality. The element of athletic asceticism which is a part of all virility, and is found even—indeed often in a high degree—among savages, has its main moral justification as one aid to sublimation. Throughout life sublimation acts by transforming some part at all events of the creative sexual energy from its elementary animal manifestations into more highly individual and social manifestations, or at all events into finer forms of sexual activity, forms that seem to us more beautiful and satisfy us more widely. Purity, we thus come to see is, in one aspect, the action of sublimation, not abolishing sexual activity, but lifting it into forms of which our best judgment may approve.

[7] We may gather the history of the term from the Oxford Dictionary. Bodies, said Davies, are transformed to spirit "by sublimation strange," and Ben Jonson in Cynthia's Revels spoke of a being "sublimated and refined"; Purchas and Jackson, early in the same seventeenth century, referred to religion as "sublimating" human nature, and Jeremy Taylor, a little later, to "subliming" marriage into a sacrament; Shaftesbury, early in the eighteenth century, spoke of human nature being "sublimated by a sort of spiritual chemists" and Welton, a little later, of "a love sublimate and refined," while, finally, and altogether in our modern sense, Peacock in 1816 in his Headlong Hall referred to "that enthusiastic sublimation which is the source of greatness and energy."

We must not suppose—as is too often assumed—that sublimation can be carried out easily, completely, or even with unmixed advantage. If it were so, certainly the old-fashioned moralist would be confronted by few difficulties, but we have ample reason to believe that it is not so. It is with sexual energy, well observes Freud, who yet attaches great importance to sublimation, as it is with heat in our machines: only a certain proportion can be transformed into work. Or, as it is put by Löwenfeld, who is not a constructive philosopher but a careful and cautious medical investigator, the advantages of sublimation are not received in specially high degree by those who permanently deny to their sexual impulse every natural direct relief. The celibate Catholic clergy, notwithstanding their heroic achievements in individual cases, can scarcely be said to display a conspicuous excess of intellectual energy, on the whole, over the non-celibate Protestant clergy; or, if we compare the English clergy before and after the Protestant Reformation, though the earlier period may reveal more daring and brilliant personages, the whole intellectual output of the later Church may claim comparison with that of the earlier Church. There are clearly other factors at work besides sublimation, and even sublimation may act most potently, not when the sexual activities sink or are driven into a tame and monotonous subordination, but rather when they assume a splendid energy which surges into many channels. Yet sublimation is a very real influence, not only in its more unconscious and profound operations, but in its more immediate and temporary applications, as part of an athletic discipline, acting best perhaps when it acts most automatically, to utilise the motor energy of the organism in the attainment of any high physical or psychic achievement.

We have to realise, however, that these transmutations do not only take place by way of a sublimation of sexual energy, but also by way of a degradation of that energy. The new form of energy produced, that is to say, may not be of a beneficial kind; it may be of a mischievous kind, a form of perversion or disease. Sexual self-denial, instead of leading to sublimation, may lead to nervous disorder when the erotic tension, failing to find a natural outlet and not sublimated to higher erotic or non-erotic ends in the real world, is transmuted into an unreal dreamland, thus undergoing what Jung terms introversion; while there are also the people already referred to, in whom immature childish sexuality persists into an adult stage of development it is no longer altogether in accord with, so that conflict, with various possible trains of nervous symptoms, may result. Disturbances and conflicts in the emotional sexual field may, we know, in these and similar ways become transformed into physical symptoms of disorder which can be seen to have a precise symbolic relationship to definite events in the patient's emotional history, while fits of nervous terror, or anxiety-neurosis, may frequently be regarded as a degradation of thwarted or disturbed sexual energy, manifesting its origin by presenting a picture of sexual excitation transposed into a non-sexual shape of an entirely useless or mischievous character.

Thus, to sum up, we may say that the sexual energy of the organism is a mighty force, automatically generated throughout life. Under healthy conditions that force is transmuted in more or less degree, but never entirely, into forms that further the development of the individual and the general ends of life. These transformations are to some extent automatic, to some extent within the control of personal guidance. But there are limits to such guidance, for the primitive human personality can never be altogether rendered an artificial creature of civilisation. When these limits are reached the transmutation of sexual energy may become useless or even dangerous, and we fail to attain the exquisite flower of Purity.

It may seem that in setting forth the nature of the sexual impulse in the light of modern biology and psychology, I have said but little of purity and less of morality. Yet that is as it should be. We must first be content to see how the machine works and watch the wheels go round. We must understand before we can pretend to control; in the natural world, as Bacon long ago said, we can only command by obeying. Moreover, in this field Nature's order is far older and more firmly established than our civilised human morality. In our arrogance we often assume that Morality is the master of Nature. Yet except when it is so elementary or fundamental as to be part of Nature, it is but a guide, and a guide that is only a child, so young, so capricious, that in every age its wayward hand has sought to pull Nature in a different direction. Even only in order to guide we must first see and know.

We realise that never more than when we observe the distinction which conventional sex-morals so often makes between men and women. Failing to find in women exactly the same kind of sexual emotions, as they find in themselves, men have concluded that there are none there at all. So man has regarded himself as the sexual animal, and woman as either the passive object of his adoring love or the helpless victim of his degrading lust, in either case as a being who, unlike man, possessed an innocent "purity" by nature, without any need for the trouble of acquiring it. Of woman as a real human being, with sexual needs and sexual responsibilities, morality has often known nothing. It has been content to preach restraint to man, an abstract and meaningless restraint even if it were possible. But when we have regard to the actual facts of life, we can no longer place virtue in a vacuum. Women are just as apt as men to be afflicted by the petty jealousies and narrownesses of the crude sexual impulse; women just as much as men need the perpetual sublimation of erotic desire into forms of more sincere purity, of larger harmony, in gaining which ends all the essential ends of morality are alone gained. The delicate adjustment of the needs of each sex to the needs of the other sex to the end of what Chaucer called fine loving, the adjustment of the needs of both sexes to the larger ends of fine living, may well furnish a perpetual moral discipline which extends its fortifying influence to men and women alike.

It is this universality of sexual emotion, blending in its own mighty stream, as is now realised, many other currents of emotion, even the parental and the filial, and traceable even in childhood,—the wide efflorescence of an energy constantly generated by a vital internal mechanism,—which renders vain all attempts either to suppress or to ignore the problem of sex, however immensely urgent we might foolishly imagine such attempts to be. Even the history of the early Christian ascetics in Egypt, as recorded in the contemporary Paradise of Palladius, illustrates the futility of seeking to quench the unquenchable, the flame of fire which is life itself. These "athletes of the Lord" were under the best possible conditions for the conquest of lust; they had been driven into the solitude of the desert by a genuine deeply-felt impulse, they could regulate their lives as they would, and they possessed an almost inconceivable energy of resolution. They were prepared to live on herbs, even to eat grass, and to undertake any labour of self-denial. They were so scrupulous that we hear of a holy man who would even efface a woman's footprints in the sand lest a brother might thereby be led into thoughts of evil. Yet they were perpetually tempted to seductive visions and desires, even after a monastic life of forty years, and the women seem to have been not less liable to yield to temptation than the men.

It may be noted that in the most perfect saints there has not always been a complete suppression of the sexual impulse even on the normal plane, nor even, in some cases, the attempt at such complete suppression. In the early days of Christianity the exercise of chastity was frequently combined with a close and romantic intimacy of affection between the sexes which shocked austere moralists. Even in the eleventh century we find that the charming and saintly Robert of Arbrissel, founder of the order of Fontevrault, would often sleep with his nuns, notwithstanding the remonstrances of pious friends who thought he was displaying too heroic a manifestation of continence, failing to understand that he was effecting a sweet compromise with continence. If, moreover, we consider the rarest and finest of the saints we usually find that in their early lives there was a period of full expansion of the organic activities in which all the natural impulses had full play. This was the case with the two greatest and most influential saints of the Christian Church, St. Augustine and St. Francis of Assisi, absolutely unlike as they were in most other respects. Sublimation, we see again and again, is limited, and the best developments of the spiritual life are not likely to come about by the rigid attempt to obtain a complete transmutation of sexual energy.

The old notion that any strict attempt to adhere to sexual abstinence is beset by terrible risks, insanity and so forth, has no foundation, at all events where we are concerned with reasonably sound and healthy people. But it is a very serious error to suppose that the effort to achieve complete and prolonged sexual abstinence is without any bad results at all, physical or psychic, either in men or women who are normal and healthy. This is now generally recognised everywhere, except in the English-speaking countries, where the supposed interests of a prudish morality often lead to a refusal to look facts in the face. As Professor Näcke, a careful and cautious physician, stated shortly before his death, a few years ago, the opinion that sexual abstinence has no bad effects is not to-day held by a single authority on questions of sex; the fight is only concerned with the nature and degree of the bad effects which, in Näcke's belief—and he was doubtless right—are never of a gravely serious character.

Yet we have also to remember that not only, as we have seen, is the effort to achieve complete abstinence—which we ignorantly term "purity"—futile, since we are concerned with a force which is being constantly generated within the organism, but in the effort to achieve it we are abusing a great source of beneficent energy. We lose more than half of what we might gain when we cover it up, and try to push it back, to produce, it may be, not harmonious activity in the world, but merely internal confusion and distortion, and perhaps the paralysis of half the soul's energy. The sexual activities of the organism, we cannot too often repeat, constitute a mighty source of energy which we can never altogether repress though by wise guidance we may render it an aid not only to personal development and well-being but to the moral betterment of the world. The attraction of sex, according to a superstition which reaches far back into antiquity, is a baleful comet pointing to destruction, rather than a mighty star to which we may harness our chariot. It may certainly be either, and which it is likely to become depends largely on our knowledge and our power of self-guidance.

In old days when, as we have seen, tradition, aided by the most fantastic superstitions, insisted on the baleful aspects of sex, the whole emphasis was placed against passion. Since knowledge and self-guidance, without which passion is likely to be in fact pernicious, were then usually absent, the emphasis was needed, and when Böhme, the old mystic, declared that the art of living is to "harness our fiery energies to the service of the light," it has recently been even maintained that he was the solitary pioneer of our modern doctrines. But the ages in which ill-regulated passion exceeded—ages at least full of vitality and energy—gave place to a more anæmic society. To-day the conditions are changed, even reversed. Moral maxims that were wholesome in feudal days are deadly now. We are in no danger of suffering from too much vitality, from too much energy in the explosive splendour of our social life. We possess, moreover, knowledge in plenty and self-restraint in plenty, even in excess, however wrongly they may sometimes be applied. It is passion, more passion and fuller, that we need. The moralist who bans passion is not of our time; his place these many years is with the dead. For we know what happens in a world when those who ban passion have triumphed. When Love is suppressed Hate takes its place. The least regulated orgies of Love grow innocent beside the orgies of Hate. When nations that might well worship one another cut one another's throats, when Cruelty and Self-righteousness and Lying and Injustice and all the Powers of Destruction rule the human heart, the world is devastated, the fibre of the whole organism, of society grows flaccid, and all the ideals of civilisation are debased. If the world is not now sick of Hate we may be sure it never will be; so whatever may happen to the world let us remember that the individual is still left, to carry on the tasks of Love, to do good even in an evil world.

It is more passion and ever more that we need if we are to undo the work of Hate, if we are to add to the gaiety and splendour of life, to the sum of human achievement, to the aspiration of human ecstasy. The things that fill men and women with beauty and exhilaration, and spur them to actions beyond themselves, are the things that are now needed. The entire intrinsic purification of the soul, it was held by the great Spanish Jesuit theologian, Suarez, takes place at the moment when, provided the soul is of good disposition, it sees God; he meant after death, but for us the saying is symbolic of the living truth. It is only in the passion of facing the naked beauty of the world and its naked truth that we can win intrinsic purity. Not all, indeed, who look upon the face of God can live. It is not well that they should live. It is only the metals that can be welded in the fire of passion to finer services that the world needs. It would be well that the rest should be lost in those flames. That indeed were a world fit to perish, wherein the moralist had set up the ignoble maxim: Safety first.

THE OBJECTS OF MARRIAGE

What are the legitimate objects of marriage? We know that many people seek to marry for ends that can scarcely be called legitimate, that men may marry to obtain a cheap domestic drudge or nurse, and that women may marry to be kept when they are tired of keeping themselves. These objects in marriage may or may not be moral, but in any case they are scarcely its legitimate ends. We are here concerned to ascertain those ends of marriage which are legitimate when we take the highest ground as moral and civilised men and women living in an advanced state of society and seeking, if we can, to advance that state of society still further.

The primary end of marriage is to beget and bear offspring, and to rear them until they are able to take care of themselves. On that basis Man is at one with all the mammals and most of the birds. If, indeed, we disregard the originally less essential part of this end—that is to say, the care and tending of the young—this end of marriage is not only the primary but usually the sole end of sexual intercourse in the whole mammal world. As a natural instinct, its achievement involves gratification and well-being, but this bait of gratification is merely a device of Nature's and not in itself an end having any useful function at the periods when conception is not possible. This is clearly indicated by the fact that among animals the female only experiences sexual desire at the season of impregnation, and that desire ceases as soon as impregnation takes place, though this is only in a few species true of the male, obviously because, if his sexual desire and aptitude were confined to so brief a period, the chances of the female meeting the right male at the right moment would be too seriously diminished; so that the attentive and inquisitive attitude towards the female by the male animal—which we may often think we see still traceable in the human species—is not the outcome of lustfulness for personal gratification ("wantonly to satisfy carnal lusts and appetites like brute beasts," as the Anglican Prayer Book incorrectly puts it) but implanted by Nature for the benefit of the female and the attainment of the primary object of procreation. This primary object we may term the animal end of marriage.

This object remains not only the primary but even the sole end of marriage among the lower races of mankind generally. The erotic idea, in its deeper sense, that is to say the element of love, arose very slowly in mankind. It is found, it is true, among some lower races, and it appears that some tribes possess a word for the joy of love in a purely psychic sense. But even among European races the evolution was late. The Greek poets, except the latest, showed little recognition of love as an element of marriage. Theognis compared marriage with cattle-breeding. The Romans of the Republic took much the same view. Greeks and Romans alike regarded breeding as the one recognisable object of marriage; any other object was mere wantonness and had better, they thought, be carried on outside marriage. Religion, which preserves so many ancient and primitive conceptions of life, has consecrated this conception also, and Christianity—though, as I will point out later, it has tended to enlarge the conception—at the outset only offered the choice between celibacy on the one hand and on the other marriage for the production of offspring.

Yet, from, an early period in human history, a secondary function of sexual intercourse had been slowly growing up to become one of the great objects of marriage. Among animals, it may be said, and even sometimes in man, the sexual impulse, when once aroused, makes but a short and swift circuit through the brain to reach its consummation. But as the brain and its faculties develop, powerfully aided indeed by the very difficulties of the sexual life, the impulse for sexual union has to traverse ever longer, slower, more painful paths, before it reaches—and sometimes it never reaches—its ultimate object. This means that sex gradually becomes intertwined with all the highest and subtlest human emotions and activities, with the refinements of social intercourse, with high adventure in every sphere, with art, with religion. The primitive animal instinct, having the sole end of procreation, becomes on its way to that end the inspiring stimulus to all those psychic energies which in civilisation we count most precious. This function is thus, we see, a by-product. But, as we know, even in our human factories, the by-product is sometimes more valuable than the product. That is so as regards the functional products of human evolution. The hand was produced out of the animal forelimb with the primary end of grasping the things we materially need, but as a by-product the hand has developed the function of making and playing the piano and the violin, and that secondary functional by-product of the hand we account, even as measured by the rough test of money, more precious, however less materially necessary, than its primary function. It is, however, only in rare and gifted natures that transformed sexual energy becomes of supreme value for its own sake without ever attaining the normal physical outlet. For the most part the by-product accompanies the product, throughout, thus adding a secondary, yet peculiarly sacred and specially human, object of marriage to its primary animal object. This may be termed the spiritual object of marriage.

By the term "spiritual" we are not to understand any mysterious and supernatural qualities. It is simply a convenient name, in distinction from animal, to cover all those higher mental and emotional processes which in human evolution are ever gaining greater power. It is needless to enumerate the constituents of this spiritual end of sexual intercourse, for everyone is entitled to enumerate them differently and in different order. They include not only all that makes love a gracious and beautiful erotic art, but the whole element of pleasure in so far as pleasure is more than a mere animal gratification. Our ancient ascetic traditions often make us blind to the meaning of pleasure. We see only its possibilities of evil and not its mightiness for good. We forget that, as Romain Rolland says, "Joy is as holy as Pain." No one has insisted so much on the supreme importance of the element of pleasure in the spiritual ends of sex as James Hinton. Rightly used, he declares, Pleasure is "the Child of God," to be recognised as a "mighty storehouse of force," and he pointed out the significant fact that in the course of human progress its importance increases rather than diminishes.[8] While it is perfectly true that sexual energy may be in large degree arrested, and transformed into intellectual and moral forms, yet it is also true that pleasure itself, and above all, sexual pleasure, wisely used and not abused, may prove the stimulus and liberator of our finest and most exalted activities. It is largely this remarkable function of sexual pleasure which is decisive in settling the argument of those who claim that continence is the only alternative to the animal end of marriage. That argument ignores the liberating and harmonising influences, giving wholesome balance and sanity to the whole organism, imparted by a sexual union which is the outcome of the psychic as well as physical needs. There is, further, in the attainment of the spiritual end of marriage, much more than the benefit of each individual separately. There is, that is to say, the effect on the union itself. For through harmonious sex relationships a deeper spiritual unity is reached than can possibly be derived from continence in or out of marriage, and the marriage association becomes an apter instrument in the service of the world. Apart from any sexual craving, the complete spiritual contact of two persons who love each other can only be attained through some act of rare intimacy. No act can be quite so intimate as the sexual embrace. In its accomplishment, for all who have reached a reasonably human degree of development, the communion of bodies becomes the communion of souls. The outward and visible sign has been the consummation of an inward and spiritual grace. "I would base all my sex teaching to children and young people on the beauty and sacredness of sex," wrote a distinguished woman; "sex intercourse is the great sacrament of life, he that eateth and drinketh unworthily eateth and drinketh his own damnation; but it may be the most beautiful sacrament between two souls who have no thought of children."[9] To many the idea of a sacrament seems merely ecclesiastical, but that is a misunderstanding. The word "sacrament" is the ancient Roman name of a soldier's oath of military allegiance, and the idea, in the deeper sense, existed long before Christianity, and has ever been regarded as the physical sign of the closest possible union with some great spiritual reality. From our modern standpoint we may say, with James Hinton, that the sexual embrace, worthily understood, can only be compared with music and with prayer. "Every true lover," it has been well said by a woman, "knows this, and the worth of any and every relationship can be judged by its success in reaching, or failing to reach, this standpoint."[10]

[8] Mrs. Havelock Ellis, James Hinton: A Sketch, Ch. IV.

[9] Olive Schreiner in a personal letter.

[10] Mrs. Havelock Ellis, James Hinton, p. 180.

I have mentioned how the Church—in part influenced by that clinging to primitive conceptions which always marks religions and in part by its ancient traditions of asceticism—tended to insist mainly, if not exclusively, on the animal object of marriage. It sought to reduce sex to a minimum because the pagans magnified sex; it banned pleasure because the Christian's path on earth was the way of the Cross; and even if theologians accepted the idea of a "Sacrament of Nature" they could only allow it to operate when the active interference of the priest was impossible, though it must in justice be said that, before the Council of Trent, the Western Church recognised that the sacrament of marriage was effected entirely by the act of the two celebrants themselves and not by the priest. Gradually, however, a more reasonable and humane opinion crept into the Church. Intercourse outside the animal end of marriage was indeed a sin, but it became merely a venial sin. The great influence of St. Augustine was on the side of allowing much freedom to intercourse outside the aim of procreation. At the Reformation, John à Lasco, a Catholic Bishop who became a Protestant and settled in England, laid it down, following various earlier theologians, that the object of marriage, besides offspring, was to serve as a "sacrament of consolation" to the united couple, and that view was more or less accepted by the founders of the Protestant churches. It is the generally accepted Protestant view to-day.[11] The importance of the spiritual end of intercourse in marriage, alike for the higher development of each member of the couple and for the intimacy and stability of their union, is still more emphatically set forth by the more advanced thinkers of to-day.

[11] It is well set forth by the Rev. H. Northcote in his excellent book, Christianity and Sex Problems.

There is something pathetic in the spectacle of those among us who are still only able to recognise the animal end of marriage, and who point to the example of the lower animals—among whom the biological conditions are entirely different—as worthy of our imitation. It has taken God—or Nature, if we will—unknown millions of years of painful struggle to evolve Man, and to raise the human species above that helpless bondage to reproduction which marks the lower animals. But on these people it has all been wasted. They are at the animal stage still. They have yet to learn the A.B.C. of love. A representative of these people in the person of an Anglican bishop, the Bishop of Southwark, appeared as a witness before the National Birth-Rate Commission which, a few years ago, met in London to investigate the decline of the birth-rate. He declared that procreation is the sole legitimate object of marriage and that intercourse for any other end was a degrading act of mere "self-gratification." This declaration had the interesting result of evoking the comments of many members of the Commission, formed of representative men and women with various stand-points—Protestant, Catholic, and other—and it is notable that while not one identified himself with the Bishop's opinion, several decisively opposed that opinion, as contrary to the best beliefs of both ancient and modern times, as representing a low and not a high moral standpoint, and as involving the notion that the whole sexual activity of an individual should be reduced to perhaps two or three effective acts of intercourse in a lifetime. Such a notion obviously cannot be carried into general practice, putting aside the question as to whether it would be desirable, and it may be added that it would have the further result of shutting out from the life of love altogether all those persons who, for whatever reason, feel that it is their duty to refrain from having children at all. It is the attitude of a handful of Pharisees seeking to thrust the bulk of mankind into Hell. All this confusion and evil comes of the blindness which cannot know that, beyond the primary animal end of propagation in marriage, there is a secondary but more exalted spiritual end.

It is needless to insist how intimately that secondary end of marriage is bound up with the practice of birth-control. Without birth-control, indeed, it could frequently have no existence at all, and even at the best seldom be free from disconcerting possibilities fatal to its very essence. Against these disconcerting possibilities is often placed, on the other side, the un-æsthetic nature of the contraceptives associated with birth-control. Yet, it must be remembered, they are of a part with the whole of our civilised human life. We at no point enter the spiritual save through the material. Forel has in this connection compared the use of contraceptives to the use of eye-glasses. Eye-glasses are equally un-æsthetic, yet they are devices, based on

Nature, wherewith to supplement the deficiencies of Nature. However in themselves un-æsthetic, for those who need them they make the æsthetic possible. Eye-glasses and contraceptives alike are a portal to the spiritual world for many who, without them, would find that world largely a closed book.

Birth-control is effecting, and promising to effect, many functions in our social life. By furnishing the means to limit the size of families, which would otherwise be excessive, it confers the greatest benefit on the family and especially on the mother. By rendering easily possible a selection in parentage and the choice of the right time and circumstances for conception it is, again, the chief key to the eugenic improvement of the race. There are many other benefits, as is now generally becoming clear, which will be derived from the rightly applied practice of birth-control. To many of us it is not the least of these that birth-control effects finally the complete liberation of the spiritual object of marriage.

HUSBANDS AND WIVES

It has always been common to discuss the psychology of women. The psychology of men has usually been passed over, whether because it is too simple or too complicated. But the marriage question to-day is much less the wife-problem than the husband-problem. Women in their personal and social activities have been slowly expanding along lines which are now generally accepted. But there has been no marked change of responsive character in the activities of men. Hence a defective adjustment of men and women, felt in all sorts of subtle as well as grosser ways, most felt when they are husband and wife, and sometimes becoming acute.

It is necessary to make clear that, as is here assumed at the outset, "man" and "husband" are not quite the same thing, even when they refer to the same person. No doubt that is also true of "woman" and "wife." A woman in her quality as woman may be a different kind of person from what she is in her function as wife. But in the case of a man the distinction is more marked. One may know a man well in the world as a man and not know him at all in his home as a husband; not necessarily that he is unfavourably revealed in the latter capacity. It is simply that he is different.

The explanation is not really far to seek. A man in the world is in vital response to the influences around him. But a husband in the home is playing a part which was created for him long centuries before he was born. He is falling into a convention, which, indeed, was moulded to fit many masculine human needs but has become rigidly traditionalised. Thus the part no longer corresponds accurately to the player's nature nor to the circumstances under which it has to be played.

In the marriage system which has prevailed in our world for several thousand years, a certain hierarchy, or sacred order in authority, has throughout been recognised. The family has been regarded as a small State of which the husband and father is head. Classic paganism and Christianity differed on many points, but they were completely at one on this. The Roman system was on a patriarchal basis and continued to be so theoretically even when in practise it came to allow great independence to the wife. Christianity, although it allowed complete spiritual freedom to the individual, introduced no fundamentally new theory of the family, and, indeed, re-inforced the old theory by regarding the family as a little church of which the husband was the head. Just as Christ is the head of the Church, St. Paul repeatedly asserted, so the husband is the head of the wife; therefore, as it was constantly argued during the Middle Ages, a man is bound to rule his wife. St. Augustine, the most influential of Christian Fathers, even said that a wife should be proud to consider herself as the servant of her husband, his ancilla, a word that had in it the suggestion of slave. That was the underlying assumption throughout the Middle Ages, for the Northern Germanic peoples, having always been accustomed to wife-purchase before their conversion, had found it quite easy to assimilate the Christian view. Protestantism, even Puritanism with its associations of spiritual revolt, so far from modifying the accepted attitude, strengthened it, for they found authority for all social organisation in the Bible, and the Bible revealed an emphatic predominance of the Jewish husband, who possessed essential rights to which the wife had no claim. Milton, who had the poet's sensitiveness to the loveliness of woman, and the lonely man's feeling for the solace of her society, was yet firmly assured of the husband's superiority over his wife. He has indeed furnished the classical picture of it in Adam and Eve,

"He for God only, she for God in him,"

and to that God she owed "subjection," even though she might qualify it by "sweet reluctant amorous delay." This was completely in harmony with the legal position of the wife. As a subject she was naturally in subjection; she owed her husband the same loyalty as a subject owes the sovereign; her disloyalty to him was termed a minor form of treason; if she murdered him the crime was legally worse than murder and she rendered herself liable to be burnt.

We see that all the influences on our civilisation, religious and secular, southern and northern, have combined to mould the underlying bony structure of our family system in such a way that, however it may appear softened and disguised on the surface, the husband is the head and the wife subject to him. We must not be supposed hereby to deny that the wife has had much authority, many privileges, considerable freedom, and in individual cases much opportunity to domineer, whatever superiority custom or brute strength may have given the husband. There are henpecked husbands, it has been remarked, even in aboriginal Australia. It is necessary to avoid the error of those enthusiasts for the emancipation of women who, out of their eager faith in the future of women, used to describe her past as one of scarcely mitigated servitude and hardship. If women had not constantly succeeded in overcoming or eluding the difficulties that beset them in the past, it would be foolish to cherish any faith in their future. It must, moreover, be remembered that the very constitution of that ecclesiastico-feudal hierarchy which made the husband supreme over the wife, also made the wife jointly with her husband supreme over their children and over their servants. The Middle Ages, alike in England and in France, as doubtless in Christendom generally, accepted the rule laid down in Gratian's Decretum, the great mediæval text-book of Canon Law, that "the husband may chastise his wife temperately, for she is of his household," but the wife might chastise her daughters and her servants, and she sometimes exercised that right in ways that we should nowadays think scarcely temperate.

If we seek to observe how the system worked some five hundred years ago when it had not yet become, as it is to-day, both weakened and disguised, we cannot do better than turn to the Paston Letters, the most instructive documents we possess concerning the domestic life of excellent yet fairly average people of the upper middle class in England in the fifteenth century. Marriage was still frankly and fundamentally (as it was in the following century and less frankly later) a commercial transaction. The wooer, when he had a wife in view, stated as a matter of course that he proposed to "deal" in the matter; it was quite recognised on both sides that love and courtship must depend on whether the "deal" came off satisfactorily. John Paston approached Sir Thomas Brews, through a third person, with a view to negotiate a marriage with his daughter Margery. She was willing, even eager, and while the matter was still uncertain she wrote him a letter on Valentine's Day, addressing him as "Right reverent and worshipful and my right well-beloved Valentine," to tell him that it was impossible for her father to offer a larger dowry than he had already promised. "If that you could be content with that good, and my poor person, I would be the merriest maiden on ground." In his first letter— boldly written, he says, without her knowledge or license—he addresses her simply as "Mistress," and assures her that "I am and will be yours and at your commandment in every wise during my life." A few weeks later, addressing him as "Right worshipful master," she calls him "mine own sweetheart," and ends up, as she frequently does, "your servant and bedeswoman." Some months later, a few weeks after marriage, she addresses her husband in the correct manner of the time as "Right reverent and worshipful husband," asking him to buy her a gown as she is weary of wearing her present one, it is so cumbrous. Five years later she refers to "all" the babies, and writes in haste: "Right reverent and worshipful Sir, in my most humble wise I recommend me unto you as lowly as I can," etc., though she adds in a postscript: "Please you to send for me for I think long since I lay in your arms." If we turn to another wife of the Paston family, a little earlier in the century, Margaret Paston, whose husband's name also was John, we find the same attitude even more distinctly expressed. She always addressed him in her most familiar letters, showing affectionate concern for his welfare, as "Right reverent and worshipful husband" or "Right worshipful master." It is seldom that he writes to her at all, but when he writes the superscription is simply "To my mistress Paston," or "my cousin," with little greeting at either beginning or end. Once only, with unexampled effusion, he writes to her as "My own dear sovereign lady" and signs himself "Your true and trusting husband."[12]

[12] We see just the same formulas in the fifteenth century letters of the Stonor family (Stonor Letters and Papers, Camden Society), though in these letters we seem often to find a lighter and more playful touch than was common among the Pastons. I may refer here to Dr. Powell's learned and well written book (with which I was not acquainted when I wrote this chapter), English Domestic Relations 1487-1653 (Columbia University Press).

If we turn to France the relation of the wife to her husband was the same, or even more definitely dependent, for he occupied the place of father to her as well as of husband and sovereign, in this respect carrying on a tradition of Roman Law. She was her husband's "wife and subject"; she signed herself "Vostre humble obéissante fille et amye." If also we turn to the Book of the Chevalier de la Tour-Landry in Anjou, written at the end of the fourteenth century, we find a picture of the relations of women to men in marriage comparable to that presented in the Paston Letters, though of a different order. This book was, as we know, written for the instruction of his daughters by a Knight who seems to have been a fairly average man of his time in his beliefs, and in character, as he has been described, probably above it, "a man of the world, a Christian, a parent, and a gentleman." His book is full of interesting light on the customs and manners of his day, though it is mainly a picture of what the writer thought ought to be rather than what always was. Herein the Knight is sagacious and moderate, much of his advice is admirably sound for every age. He is less concerned with affirming the authority of husbands than with assuring the happiness and well-being of his dearly loved daughters. But he clearly finds this bound up with the recognition of the authority of the husband, and the demands he makes are fairly concordant with the relationships we see established among the Pastons. The Knight abounds in illustrations, from Lot's daughters down to his own time, for the example or the warning of his daughters. The ideal he holds up to them is strictly domestic and in a sense conventional. He puts the matter on practical rather than religious or legal grounds, and his fundamental assumption is "that no woman ought ever to thwart or refuse to obey the ordinance of her lord; that is, if she is either desirous to be mistress of his affections or to have peace and understanding in the house. For very evident reasons submission should begin on her part." One would like to know what duties the Knight inculcated on husbands, but the corresponding book he wrote for the guidance of his sons appears no longer to be extant.

On the whole, the fundamental traditions of our western world concerning the duties of husbands and wives are well summed up in what Pollock and Maitland term "that curious cabinet of antiquities, the marriage ritual of the English Church." Here we find that the husband promises to love and cherish the wife, but she promises not only to love and cherish but also to obey him, though, it may be noted, this point was not introduced into English marriage rites until the fourteenth century, when the wife promised to be "buxom" (which then meant submissive) and "bonair" (courteous and kind), while in some French and Spanish rites it has never been introduced at all. But we may take it to be generally implied. In the final address to the married couple the priest admonishes the bride that the husband is the head of the wife, and that her part is submission. In some more ancient and local rituals this point was further driven home, and on the delivery of the ring the bride knelt and kissed the bridegroom's right foot. In course of time this was modified, at all events in France, and she simply dropped the ring, so that her motion of stooping was regarded as for the purpose of picking it up. I note that change for it is significant of the ways in which we modify the traditions of the past, not quite abandoning them but pretending that they have other than the fundamental original motives. We see just the same thing in the use of the ring, which was in the first place a part of the bride-price, frequently accompanied by money, proof that the wife had been duly purchased. It was thus made easy to regard the ring as really a golden fetter. That idea soon became offensive, and the new idea was originated that the ring was a pledge of affection; thus, quite early in some countries, the husband, also wore a wedding ring.

The marriage order illustrated by the Paston Letters and the Book of the Chevalier de la Tour-Landry before the Reformation, and the Anglican Book of Common Prayer afterwards, has never been definitely broken; it is a part of our living tradition to-day. But during recent centuries it has been overlaid by the growth of new fashions and sentiments which have softened its hard outlines to the view. It has been disguised, notably during the eighteenth century, by the development of a new feeling of social equality, chiefly initiated in France, which, in an atmosphere of public intercourse largely regulated by women, made the ostentatious assertion of the husband's headship over his wife displeasing and even ridiculous. Then, especially in the nineteenth century, there began another movement, chiefly initiated in England and carried further in America, which affected the foundations of the husband's position from beneath. This movement consisted in a great number of legislative measures and judicial pronouncements and administrative orders—each small in itself and never co-ordinated—which taken altogether have had a cumulative effect in immensely increasing the rights of the wife independently of her husband or even in opposition to him. Thus at the present time the husband's authority has been overlaid by new social conventions from above and undermined by new legal regulations from below.

Yet, it is important to realise, although the husband's domestic throne has been in appearance elegantly re-covered and in substance has become worm-eaten, it still stands and still retains its ancient shape and structure. There has never been a French Revolution in the home, and that Revolution itself, which modified society so extensively, scarcely modified the legal supremacy of the husband at all, even in France under the Code Napoléon and still less anywhere else. Interwoven with all the new developments, and however less obtrusive it may have become, the old tradition still continues among us. Since, also, the husband is, conventionally and in large measure really, the economic support of the home,—the work of the wife and even actual financial contributions brought by her not being supposed to affect that convention,—this state of things is held to be justified.

Thus when a man enters the home as a husband, to seat himself on the antique domestic throne and to play the part assigned to him of old, he is involuntarily, even unconsciously, following an ancient tradition and taking his place in a procession of husbands which began long ages before he was born. It thus comes about that a man, even after he is married, and a husband are two different persons, so that his wife who mainly knows him as a husband may be unable to form any just idea of what he is like as a man. As a husband he has stepped out of the path that belongs to him in the world, and taken on another part which has called out altogether different reactions, so he is sometimes a much more admirable person in one of these spheres—whichever it may be—than in the other.

We must not be surprised if the husband's position has sometimes developed those qualities which from the modern point of view are the less admirable. In this respect the sovereign husband resembles the Sovereign State. The Sovereign State, as it has survived from Renaissance days in our modern world, may be made up of admirable people, yet as a State they are forced into an attitude of helpless egoism which nowadays fails to commend itself to the outside world, and the tendency of scientific jurists to-day is to deal very critically with the old conception of the Sovereign State. It is so with the husband in the home. He was thrust by ancient tradition into a position of sovereignty which impelled him to play a part of helpless egoism. He was a celestial body in the home around which all the other inmates were revolving satellites. The hours of rising and retiring, the times of meals and their nature and substance, all the activities of the household—in which he himself takes little or no part—are still arranged primarily to suit his work, his play, and his tastes. This is an accepted matter of course, and not the result of any violent self-assertion on his part. It is equally an accepted matter of course that the wife should be constantly occupied in keeping this little solar system in easy harmonious movement, evolving from it, if she has the skill, the music of the spheres. She has no recognised independent personality of her own, nor even any right to go away by herself for a little change and recreation. Any work of her own, play of her own, tastes of her own, must be strictly subordinated, if not suppressed altogether.

In the old days, from which our domestic traditions proceed, little hardship was thus inflicted on the wife. Her rights and privileges were, indeed, far less than those of the modern woman, but for that very reason the home offered her a larger field; beneath the shelter of her husband the irresponsible wife might exert a maximum of influential activity with a minimum of rights and privileges of her own. To many men, even to-day, that state of things seems the realisation of an ideal.

Yet to women it seems increasingly less so, and of necessity since the cleavage between the position of woman in society and law, and the position of the wife in the sacramental bonds of wedlock, is daily becoming greater. To-day a woman, who possibly for ten years has been leading her own life of independent work, earning her own living, choosing her own conditions in accordance with her own needs, and selecting her own periods of recreation in accordance with her own tastes, whether or not this may have included the society of a man-friend—such a woman suddenly finds on marriage, and without any assertion of authority on her husband's part, that all the outward circumstances of her life are reversed and all her inner spontaneous movements arrested. There may be no signs of this on the surface of her conduct. She loves her husband too much to wish to hurt his feelings by explaining the situation, and she values domestic peace too much to risk friction by making unexpected claims. But beneath the surface there is often a profound discontent, and even in women who thought they had gained an insight into life, a sense of disillusion. Everyone knows this who is privileged to catch a glimpse into the hearts of women—often women of most distinguished intelligence as well as women of quite ordinary nature—who leave a life of spontaneous activity in the world to enter the home.[13]

[13] While this condition of things is sometimes to be found in the more distinguished minority and in well-to-do families, it is, of course, among the great labouring majority that it is most conspicuous. Mrs. Will Crooks, of Poplar, speaking to a newspaper reporter (Daily Chronicle, 17 Feb., 1919),

truly remarked: "At present the average married woman's working day is a flagrant contradiction of all trade-union ideals. The poor thing is slaving all the time! What she needs—what she longs for—is just a little break or change now and again, an opportunity to get her mind off her work and its worries. If her husband's hours are reduced to eight, well that gives her a chance, doesn't it? The home and the children are, after all, as much his as hers. With his enlarged leisure he will now be able to take a fair share in home duties. I suggest that they take it turn and turn about—one night he goes out and she looks after the house and the children; the next night she goes out and he takes charge of things at home. She can sometimes go to the cinema, sometimes call on friends. Then, say once a week, they can both go out together, taking the children with them. That will be a little change and treat for everybody."

It is not to be supposed that in this presentation of the situation in the home, as it is to-day visible to those who are privileged to see beneath the surface, any accusation is brought against the husband. He is no more guilty of an unreasonable conservatism than the wife is guilty of an unreasonable radicalism. Each of them is the outcome of a tradition. The point is that the events of the past hundred years have produced a discrepancy in the two lines of tradition, with a resultant lack of harmony, independent of the goodwill of either husband or wife.

Olive Schreiner, in her Woman and Labour, has eloquently set forth the tendency to parasitism which civilisation produces in women; they no longer exercise the arts and industries which were theirs in former ages, and so they become economically dependent on men, losing their energies and aptitudes, and becoming like those dull parasitic animals which live as blood-suckers of their host. That picture, which was of course never true of all women, is now ceasing to be true of any but a negligible minority; it presents, moreover, a parasitism limited to the economic side of life. For if the wife has often been a lazy gold-sucking parasite on her husband in the world, the husband has yet oftener been a helpless service-absorbing parasite on his wife in the home. There is, that is to say, not only an economic parasitism, with no adequate return for financial support, but a still more prevalent domestic parasitism, with an absorption of services for which no return would be adequate. There are many helpful husbands in the home, but there are a larger number who are helpless and have never been trained to be anything else but helpless, even by their wives, who would often detest a rival in household work and management. The average husband enjoys the total effect of his home but is usually unable to contribute any of the details of work and organisation that make it enjoyable. He cannot keep it in order and cleanliness and regulated movement, he seldom knows how to buy the things that are needed for its upkeep, nor how to prepare and cook and present a decent meal; he cannot even attend to his own domestic needs. It is the wife's consolation that most husbands are not always at home.

"In ministering to the wants of the family, the woman has reduced man to a state of considerable dependency on her in all domestic affairs, just as she is dependent on him for bodily protection. In the course of ages this has gone so far as to foster a peculiar helplessness on the part of the man, which manifests itself in a somewhat childlike reliance of the husband on the wife. In fact it may be said that the husband is, to all intents and purposes, incapable of maintaining himself without the aid of a woman." This passage will probably seem to many readers to apply quite fairly well to men as they exist to-day in most of those lands which we consider at the summit of our civilisation. Yet it was not written of civilisation, or of white men, but of the Bantu tribes of East Africa,[14] complete Negroes who, while far from being among the lowest savages, belong to a culture which is only just emerging from cannibalism, witchcraft, and customary bloodshed. So close a resemblance between the European husband and the Negro husband significantly suggests how remarkable has been the arrest of development in the husband's customary status during a vast period of the world's history.

[14] Hon. C. Dundas, Journal of the Anthropological Institute, Vol. 45, 1915, p. 302.

It is in the considerable group of couples where the husband's work separates him but little from the home that the pressure on the wife is most severe, and without the relief and variety secured by his frequent absence. She has perhaps led a life of her own before marriage, she knows how to be economically independent; now they occupy a small dwelling, they have, maybe, one or two small children, they can only afford one helper in the work or none at all, and in this busy little hive the husband and wife are constantly tumbling over each other. It is small wonder if the wife feels a deep discontent beneath her willing ministrations and misses the devotion of the lover in the perpetual claims of the husband.

But the difficulty is not settled if she persuades him to take a room outside. He is devoted to his wife and his home, with good reason, for the wife makes the home and he is incapable of making a home. His new domestic arrangements sink into careless and sordid disorder, and he is conscious of profound discomfort. His wife soon realises that it is a choice between his return to the home and complete separation. Most wives never get even as far as this attempt at solution of the difficulty and hide their secret discontent.

This is the situation which to-day is becoming intensified and extended on a vast scale. The habit and the taste for freedom, adventure, and economic independence is becoming generated among millions of women who once meekly trod the ancient beaten paths, and we must not be so foolish as to suppose that they can suddenly renounce those habits and tastes at the threshold of marriage. Moreover, it is becoming clear to men and to women alike, and for the first time, that the world can be remoulded, and that the claims for better conditions of work, for a higher standard of life, and for the attainment of leisure, which previously had only feebly been put forward, may now be asserted drastically. We see therefore to-day a great revolutionary movement, mainly on the part of men in the world of Labour, and we see a corresponding movement, however less ostentatious, mainly on the part of women, in the world of the Home.

It may seem to some that this new movement of upheaval in the sphere of the Home is merely destructive. Timid souls have felt the like in every period of transition, and with as little reason. Just as we realise that the movement now in progress in the world of Labour for a higher standard of life and for, as it has been termed, a larger "leisure-ration," represents a wholesome revolt against the crushing conditions of prolonged monotonous work—the most deadening of all work—and a real advance towards those ideals of democracy which are still so remote, so it is with the movement in the Home. That also is the claim for a new and fairer allotment of responsibility, of larger opportunities for freedom and leisure. If in the home the husband is still to be regarded as the capitalist and the wife as the labourer, then at all events it has to be recognised that he owes her not only the satisfaction of her physical needs of food and shelter and clothing, but the opportunity to satisfy the personal spontaneous claims of her own individual nature. Just as the readjustment of Labour is really only an approach to the long recognised ideals of Democracy, so the readjustment of the Home, far from being subversive or revolutionary, is merely an approximation to the long recognised ideals of marriage.

How in practice, one may finally ask, is this readjustment of the home likely to be carried out?

In the first place we are justified in believing that in the future home men will no longer be so helpless, so domestically parasitic, as in the past. This change is indeed already coming about. It is an inestimable benefit throughout life for a man to have been forcibly lifted out of the routine comforts and feminine services of the old-fashioned home and to be thrown into an alien and solitary environment, face to face with Nature and the essential domestic human needs (in my own case I owe an inestimable debt to the chance that thus flung me into the Australian bush in early life), and one may note that the Great War has had, directly and indirectly, a remarkable influence in this direction, for it not only compelled women to exercise many enlarging and fortifying functions commonly counted as pertaining to men, it also compelled men, deprived of accustomed feminine services, to develop a new independent ability for organising domesticity, and that ability, even though it is not permanently exercised in rendering domestic services, must yet always make clear the nature of domestic problems and tend to prevent the demand for unnecessary domestic services.

But there is another quite different and more general line along which we may expect this problem to be largely solved. That is by the simplification and organisation of domestic life. If that process were carried to the full extent that is now becoming possible a large part of the problem before us would be at once solved. A great promise for the future of domestic life is held out by the growing adoption of birth-control, by which the wife and mother is relieved from that burden of unduly frequent and unwanted maternity which in the past so often crushed her vitality and destroyed her freshness. But many minor agencies are helpful. To supply heat, light, and motive power even to small households, to replace the wasteful, extravagant, and often inefficient home-cookery by meals cooked outside, as well as to facilitate the growing social habit of taking meals in spacious public restaurants, under more attractive, economical, and wholesome conditions than can usually be secured within the narrow confines of the home, to contract with specially trained workers from outside for all those routines of domestic drudgery which are often so inefficiently and laboriously carried on by the household-worker, whether mistress or servant, and to seek perpetually by new devices to simplify, which often means to beautify, all the everyday processes of life—to effect this in any comprehensive degree is to transform the home from the intolerable burden it is sometimes felt to be into a possible haven of peace and joy.[15] The trouble in the past, and even to-day, has been, not in any difficulty in providing the facilities but in prevailing people to adopt them. Thus in England, even under the stress of the Great War, there was among the working population a considerable disinclination—founded on stupid conservatism and a meaningless pride—to take advantage of National Kitchens and National Restaurants, notwithstanding the superiority of the meals in quality, cheapness, and convenience, to the workers' home meals, so that many of these establishments, even while still fostered by the Government, had speedily to close their doors. Ancient traditions, that have now become not only empty but mischievous, in these matters still fetter the wife even more than the husband. We cannot regulate even the material side of life without cultivating that intelligence in the development of which civilisation so largely consists.

[15] This aspect of the future of domesticity was often set forth by Mrs. Havelock Ellis, The New Horizon in Love and Life, 1921.

Intelligence, and even something more than intelligence, is needed along the third line of progress towards the modernised home. Simplification and organisation can effect nothing in the desired transformation if they merely end in themselves. They are only helpful in so far as they economise energy, offer a more ample leisure, and extend the opportunities for that play of the intellect, that liberation of the emotions with accompanying discipline of the primitive instincts, which are needed not only for the development of civilisation in general, but in particular of the home. Domineering egotism, the assertion of greedy possessive rights, are out of place in the modern home. They are just as mischievous when exhibited by the wife as by the husband. We have seen, as we look back, the futility in the end of the ancient structure of the home, however reasonable it was at the beginning, under our different modern social conditions, and for women to attempt nowadays to reintroduce the same structure, merely reversed would be not only mischievous but silly. That spirit of narrow exclusiveness and self centred egoism—even if it were sometimes an égoisme à deux—evoked, half a century ago, the scathing sarcasm of James Hinton, who never wearied of denouncing the "virtuous and happy homes" which he saw as "floating blotches of verdure on a sea of filth." Such outbursts seem extravagant, but they were the extravagance of an idealist at the vision which, as a physician in touch with realities, he had, seen beneath the surface of the home.

It is well to insist on the organisation of the mechanical and material side of life. Some leaders of women movements feel this so strongly that they insist on nothing else. In old days it was conventionally supposed that women's sphere was that of the feelings; the result has been that women now often take ostentatious pleasure in washing their hands of feelings and accusing men of "sentiment." But that wrongly debased word stands for the whole superstructure of life on the basis of material organisation, for all the finer and higher parts of our nature, for the greater part of civilisation.[16] The elaboration of the mechanical side of life by itself may merely serve to speed up the pace of life instead of expanding leisure, to pile up the weary burden of luxury, and still further to dissipate the energy of life in petty or frivolous channels.[17] To bring order into the region of soulless machinery running at random, to raise the super-structure of a genuinely human civilisation, is not a task which either men or women can afford to fling contemptuously to the opposite sex. It concerns them both equally and can only be carried out by both equally, working side by side in the most intimate spirit of mutual comprehension, confiding trust, and the goodwill to conquer the demon of jealousy, that dragon which slays love under the pretence of keeping it alive.

[16] "The growth of the sentiments," remarks an influential psychologist of our own time (W. McDougall, Social Psychology, p. 160), "is of the utmost importance for the character and conduct of individuals and of societies; it is the organisation of the affective and conative life. In the absence of sentiments our emotional life would be a mere chaos, without order, consistency, or continuity of any kind; and all our social relations and conduct, being based on the emotions and their impulses would be

correspondingly chaotic, unpredictable, and unstable.... Again, our judgments of value and of merit are rooted in our sentiments; and our moral principles have the same source, for they are formed by our judgments of moral value."

[17] The destructive effects of the mechanisation of modern life have lately been admirably set forth, and with much precise illustration, by Dr. Austin Freeman, Social Decay and Regeneration.

This task, it may finally be added, is always an adventure. However well organised the foundations of life may be, life must always be full of risks. We may smile, therefore, when it is remarked that the future developments of the home are risky. Birds in the air and fishes in the sea, quite as much as our own ancestors on the earth, have always found life full of risks. It was the greatest risk of all when they insisted on continuing on the old outworn ways and so became extinct. If the home is an experiment and a risky experiment, one can only say that life is always like that. We have to see to it that in this central experiment, on which our happiness so largely depends, all our finest qualities are mobilised. Even the smallest homes under the new conditions cannot be built to last with small minds and small hearts. Indeed the discipline of the home demands not only the best intellectual qualities that are available, but often involves—and in men as well as in women—a spiritual training fit to make sweeter and more generous saints than any cloister. The greater the freedom, the more complete the equality of husband and wife, the greater the possibilities of discipline and development. In view of the rigidities and injustices of the law, many couples nowadays dispense with legal marriage, and form their own private contract; that method has sometimes proved more favourable to the fidelity and permanence of love than external compulsion; it assists the husband to remain the lover, and it is often the lover more than the husband that the modern woman needs; but it has always to be remembered that in the present condition of law and social opinion a slur is cast on the children of such unions. No doubt, however, marriage and the home will undergo modifications, which will tend to make these ancient institutions a little more flexible and to permit a greater degree of variation to meet special circumstances. We can occupy ourselves with no more essential task, whether as regards ourselves or the race, than to make more beautiful the House of Life for the dwelling of Love.

THE LOVE-RIGHTS OF WOMEN

What is the part of woman, one is sometimes asked, in the sex act? Must it be the wife's concern in the marital embrace to sacrifice her own wishes from a sense of love and duty towards her husband? Or is the wife entitled to an equal mutual interest and joy in this act with her husband? It seems a simple problem. In so fundamental a relationship, which goes back to the beginning of sex in the dawn of life, it might appear that we could leave Nature to decide. Yet it is not so. Throughout the history of civilisation, wherever we can trace the feelings and ideas which have prevailed on this matter and the resultant conduct, the problem has existed, often to produce discord, conflict, and misery. The problem still exists to-day and with as important results as in the past.

In Nature, before the arrival of Man, it can scarcely be said indeed that any difficulty existed. It was taken for granted at that time that the female had both the right to her own body, and the right to a certain amount of enjoyment in the use of it. It often cost the male a serious amount of trouble—though he never failed to find it worth while—to explain to her the point where he may be allowed to come in, and to persuade her that he can contribute to her enjoyment. So it generally is throughout Nature, before we reach Man, and, though it is not invariably obvious, we often find it even among the unlikeliest animals. As is well known, it is most pronounced among the birds, who have in some species carried the erotic art,—and the faithful devotion which properly accompanied the erotic art as being an essential part of it,—to the highest point. We have here the great natural fact of courtship. Throughout Nature, wherever we meet with animals of a high type, often indeed when they are of a lowly type— provided they have not been rendered unnatural by domestication—every act of sexual union is preceded by a process of courtship. There is a sound

53

physiological reason for this courtship, for in the act of wooing and being wooed the psychic excitement gradually generated in the brains of the two partners acts as a stimulant to arouse into full activity the mechanism which ensures sexual union and aids ultimate impregnation. Such courtship is thus a fundamental natural fact.

It is as a natural fact that we still find it in full development among a large number of peoples of the lower races whom we are accustomed to regard as more primitive than ourselves. New conditions, it is true, soon enter to complicate the picture presented by savage courtship. The economic element of bargaining, destined to prove so important, comes in at an early stage. And among peoples leading a violent life, and constantly fighting, it has sometimes happened, though not always, that courtship also has been violent. This is not so frequent as was once supposed. With better knowledge it was found that the seeming brutality once thought to take the place of courtship among various peoples in a low state of culture was really itself courtship, a rough kind of play agreeable to both parties and not depriving the feminine partner of her own freedom of choice. This was notably the case as regards so-called "marriage by capture." While this is sometimes a real capture, it is more often a mock capture; the lover perhaps pursues the beloved on horseback, but she is as fleet and as skilful as he is, cannot be captured unless she wishes to be captured, and in addition, as among the Kirghiz, she may be armed with a formidable whip; so that "marriage by capture," far from being a hardship imposed on women is largely a concession to their modesty and a gratification of their erotic impulses. Even when the chief part of the decision rests with masculine force courtship is still not necessarily or usually excluded, for the exhibition of force by a lover,—and this is true for civilised as well as for savage women,—is itself a source of pleasurable stimulation, and when that is so the essence of courtship may be attained even more successfully by the forceful than by the humble lover.

The evolution of society, however, tended to overlay and sometimes even to suppress those fundamental natural tendencies. The position of the man as the sole and uncontested head of the family, the insistence on paternity and male descent, the accompanying economic developments, and the tendency to view a woman less as a self-disposing individual than as an object of barter belonging to her father, the consequent rigidity of the marriage bond and the stern insistence on wifely fidelity—all these conditions of developing civilisation, while still leaving courtship possible, diminished its significance and even abolished its necessity. Moreover, on the basis of the social, economic, and legal developments thus established, new moral, spiritual, and religious forces were slowly generated, which worked on these rules of merely exterior order, and interiorised them, thus giving them power over the souls as well as over the bodies of women.

The result was that, directly and indirectly, the legal, economic, and erotic rights of women were all diminished. It is with the erotic rights only that we are here concerned.

No doubt in its erotic aspects, as well as in its legal and economic aspects, the social order thus established was described, and in good faith, as beneficial to women, and even as maintained in their interests. Monogamy and the home, it was claimed, alike existed for the benefit and protection of women. It was not so often explained that they greatly benefited and protected men, with, moreover, this additional advantage that while women were absolutely confined to the home, men were free to exercise their activities outside the home, even, with tacit general consent, on the erotic side.

Whatever the real benefits, and there is no occasion for questioning them, of the sexual order thus established, it becomes clear that in certain important respects it had an unnatural and repressive influence on the erotic aspect of woman's sexual life. It fostered the reproductive side of woman's sexual life, but it rendered difficult for her the satisfaction of the instinct for that courtship which is the natural preliminary of reproductive activity, an instinct even more highly developed in the female than in the male, and the more insistent because in the order of Nature the burden of maternity is preceded by the reward of pleasure. But the marriage order which had become established led to the indirect result of banning pleasure in women, or at all events in wives. It was regarded as too dangerous, and even as degrading. The women who wanted pleasure were not considered fit for the home, but more suited to be devoted to an exclusive "life of pleasure," which soon turned out to be not their own pleasure but men's. A "life of pleasure," in that sense or in any other sense, was not what more than a small minority of women ever desired. The desire of women for courtship is not a thing by itself, and was not implanted for gratification by itself. It is naturally intertwined—and to a much greater degree than the corresponding desire in men—with her deepest personal, family, and social instincts, so that if these are desecrated and lost its charm soon fades.

The practices and the ideals of this established morality were both due to men, and both were so thoroughly fashioned that they subjugated alike the actions and the feelings of women. There is no sphere which we regard as so peculiarly women's sphere as that of love. Yet there is no sphere which in civilisation women have so far had so small a part in regulating. Their deepest impulses—their modesty, their maternity, their devotion, their emotional receptivity—were used, with no conscious and deliberate Machiavellism, against themselves, to mould a moral world for their habitation which they would not themselves have moulded. It is not of modern creation, nor by any means due, as some have supposed, to the asceticism of Christianity, however much Christianity may have reinforced it. Indeed one may say that in course of time Christianity had an influence in weakening it, for Christianity discovered a new reservoir of tender emotion, and such emotion may be transferred, and, as a matter of fact, was transferred, from its first religious channel into erotic channels which were thereby deepened and extended, and without reference to any design of Christianity. For the ends we achieve are often by no means those which we set out to accomplish. In ancient classic days this moral order was even more severely established than in the Middle Ages. Montaigne, in the sixteenth century, declared that "marriage is a devout and religious relationship, the pleasures derived from it should be restrained and serious, mixed with some severity." But in this matter he was not merely expressing the Christian standpoint but even more that of paganism, and he thoroughly agreed with the old Greek moralist that a man should approach his wife "prudently and severely" for fear of inciting her to lasciviousness; he thought that marriage was best arranged by a third party, and was inclined to think, with the ancients, that women are not fitted to make friends of. Montaigne has elsewhere spoken with insight of women's instinctive knowledge of the art and discipline of love and has pointed out how men have imposed their own ideals and rules of action on women from whom they have demanded opposite and contradictory virtues; yet, we see, he approves of this state of things and never suggests that women have any right to opinions of their own or feelings of their own when the sacred institution of marriage is in question.

Montaigne represents the more exalted aspects of the Pagan-Christian conception of morality in marriage which still largely prevails. But that conception lent itself to deductions, frankly accepted even by Montaigne himself, which were by no means exalted. "I find," said Montaigne, "that Venus, after all, is nothing more than the pleasure of discharging our vessels, just as nature renders pleasurable the discharges from other parts." Sir Thomas More among Catholics, and Luther among Protestants, said exactly the same thing in other and even clearer words, while untold millions of husbands in Christendom down to to-day, whether or not they have had the wit to put their theory into a phrase, have regularly put it into practice, at all events within the consecrated pale of marriage, and treated their wives, "severely and prudently," as convenient utensils for the reception of a natural excretion.

Obviously, in this view of marriage, sexual activity was regarded as an exclusively masculine function, in the exercise of which women had merely a passive part to play. Any active participation on her side thus seemed unnecessary, and even unbefitting, finally, though only in comparatively modern times, disgusting and actually degrading. Thus Acton, who was regarded half a century ago as the chief English authority on sexual matters, declared that, "happily for society," the supposition that women possess sexual feelings could be put aside as "a vile aspersion," while another medical authority of the same period stated in regard to the most simple physical sign of healthy sexual emotion that it "only happens in lascivious women." This final triumph of the masculine ideals and rule of life was, however, only achieved slowly. It was the culmination of an elaborate process of training. At the outset men had found it impossible to speak too strongly of the "wantonness" of women. This attitude was pronounced among the ancient Greeks and prominent in their dramatists. Christianity again, which ended by making women into the chief pillars of the Church, began by regarding them as the "Gate of Hell." Again, later, when in the Middle Ages this masculine moral order approached the task of subjugating the barbarians of Northern Europe, men were horrified at the licentiousness of those northern women at whose coldness they are now shocked.

That, indeed, was, as Montaigne had seen, the central core of conflict in the rule of life imposed by men on woman. Men were perpetually striving, by ways the most methodical, the most subtle, the most far-reaching, to achieve a result in women, which, when achieved, men themselves viewed with dismay. They may be said to be moved in this sphere by two passions, the passion for virtue and the passion for vice. But it so happens that both these streams of passion have to be directed at the same fascinating object: Woman. No doubt nothing is more admirable than the skill with which women have acquired the duplicity necessary to play the two contradictory parts thus imposed upon them. But in that requirement the play of their natural reactions tended to become paralysed, and the delicate mechanism of their instincts often disturbed. They were forbidden, except in a few carefully etiquetted forms, the free play of courtship, without which they could not perform their part in the erotic life with full satisfaction either to themselves or their partners. They were reduced to an artificial simulation of coldness or of warmth, according to the particular stage of the dominating masculine ideal of woman which their partner chanced to have reached. But that is an attitude equally unsatisfactory to themselves and to their lovers, even when the latter have not sufficient insight to see through its unreality. It is an attitude so unnatural and artificial that it inevitably tends to produce a real coldness which nothing can disguise. It is true that women whose instincts are not perverted at the roots do not desire to be cold. Far from it. But to dispel that coldness the right atmosphere is needed, and the insight and skill of the right man. In the erotic sphere a woman asks nothing better of a man than to be lifted above her coldness, to the higher plane where there is reciprocal interest and mutual joy in the act of love. Therein her silent demand is one with Nature's. For the biological order of the world involves those claims which, in the human range, are the erotic rights of women.

The social claims of women, their economic claims, their political claims, have long been before the world. Women themselves have actively asserted them, and they are all in process of realisation. The erotic claims of women, which are at least as fundamental, are not publicly voiced, and women themselves would be the last to assert them. It is easy to understand why that should be so. The natural and acquired qualities of women, even the qualities developed in the art of courtship, have all been utilised in building up the masculine ideal of sexual morality; it is on feminine characteristics that this masculine ideal has been based, so that women have been helpless to protest against it. Moreover, even if that were not so, to formulate such rights is to raise the question whether there so much as exists anything that can be called "erotic rights." The right to joy cannot be claimed in the same way as one claims the right to put a voting paper in a ballot box. A human being's erotic aptitudes can only be developed where the right atmosphere for them exists, and where the attitudes of both persons concerned are in harmonious sympathy. That is why the erotic rights of women have been the last of all to be attained.

Yet to-day we see a change here. The change required is, it has been said, a change of attitude and a resultant change in the atmosphere in which the sexual impulses are manifested. It involves no necessary change in the external order of our marriage system, for, as has already been pointed out, it was a coincident and not designed part of that order. Various recent lines of tendency have converged to produce this change of attitude and of atmosphere. In part the men of to-day are far more ready than the men of former days to look upon women as their comrades in the every day work of the world, instead of as beings who were ideally on a level above themselves and practically on a level considerably below themselves. In part there is the growing recognition that women have conquered many elementary human rights of which before they were deprived, and are more and more taking the position of citizens, with the same kinds of duties, privileges, and responsibilities as men. In part, also, it may be added, there is a growing diffusion among educated people of a knowledge of the primary facts of life in the two sexes, slowly dissipating and dissolving many foolish and often mischievous superstitions. The result is that, as many competent observers have noted, the young men of to-day show a new attitude towards women and towards marriage, an attitude of simplicity and frankness, a desire for mutual confidence, a readiness to discuss difficulties, an appeal to understand and to be understood. Such an attitude, which had hitherto been hard to attain, at once creates the atmosphere in which alone the free spontaneous erotic activities of women can breathe and live.

This consummation, we have seen, may be regarded as the attainment of certain rights, the corollary of other rights in the social field which women are slowly achieving as human beings on the same human level as men. It opens to women, on whom is always laid the chief burden of sex, the right to the joy and exaltation of sex, to the uplifting of the soul which, when the right conditions are fulfilled, is the outcome of the intimate approach and union of two human beings. Yet while we may find convenient so to formulate it, we need to remember that that is only a fashion of speech, for there are no rights in Nature. If we take a broader sweep, what we may choose to call an erotic right is simply the perfect poise of the conflicting forces of life, the rhythmic harmony in which generation is achieved with the highest degree of perfection compatible with the make of the world. It is our part to transform Nature's large conception into our own smaller organic mould, not otherwise than the plants, to whom we are far back akin, who dig their flexible roots deep into the moist and fruitful earth, and so are able to lift up glorious heads toward the sky.

THE PLAY-FUNCTION OF SEX

When we hear the sexual functions spoken of we commonly understand the performance of an act which normally tends to the propagation of the race. When we see the question of sexual abstinence discussed, when the desirability of sexual gratification is asserted or denied, when the idea arises of the erotic rights and needs of woman, it is always the same act with its physical results that is chiefly in mind. Such a conception is quite adequate for practical working purposes in the social world. It enables us to deal with all our established human institutions in the sphere of sex, as the arbitrary assumptions of Euclid enable us to traverse the field of elementary geometry. But beyond these useful purposes it is inadequate and even inexact. The functions of sex on the psychic and erotic side are of far greater extension than any act of procreation, they may even exclude it altogether, and when we are concerned with the welfare of the individual human being we must enlarge our outlook and deepen our insight.

There are, we know, two main functions in the sexual relationship, or what in the biological sense we term "marriage," among civilised human beings, the primary physiological function of begetting and bearing offspring and the secondary spiritual function of furthering the higher mental and emotional processes. These are the main functions of the sexual impulse, and in order to understand any further object of the sexual relationship—or even in order to understand all that is involved in the secondary object of marriage—we must go beyond conscious motives and consider the nature of the sexual impulse, physical and psychic, as rooted in the human organism.

The human organism, as we know, is a machine on which excitations from without, streaming through the nerves and brain, effect internal work, and, notably, stimulate the glandular system. In recent years the glandular

61

system, and especially that of the ductless glands, has taken on an altogether new significance. These ductless glands, as we know, liberate into the blood what are termed "hormones," or chemical messengers, which have a complex but precise action in exciting and developing all those physical and psychic activities which make up a full life alike on the general side and the reproductive side, so that their balanced functions are essential to wholesome and complete existence. In a rudimentary form these functions may be traced back to our earliest ancestors who possessed brains. In those times the predominant sense for arousing the internal mental and emotional faculties was that of smell, the other senses being gradually evolved subsequently, and it is significant that the pituitary, one of the chief ductless glands active in ourselves to-day, was developed out of the nervous centre for smell in conjunction with the membrane of the mouth. The energies of the whole organism were set in action through stimuli arising from the outside world by way of the sense of smell. In process of time the mechanism has become immensely elaborated, yet its healthy activity is ultimately dependent on a rich and varied action and reaction with the external world. It is becoming recognised that the tendency to pluri-glandular insufficiency, with its resulting lack of organic harmony and equilibrium, can be counteracted by the physical and psychic stimuli of intimate contacts with the external world. In this action and reaction, moreover, we cannot distinguish between sexual ends and general ends. The activities of the ductless glands and their hormones equally serve both ends in ways that cannot be distinguished. "The individual metabolism," as a distinguished authority in this field has expressed it, "is the reproductive metabolism."[18] Thus the establishment of our complete activities as human beings in the world is aided by, if not indeed ultimately dependent upon, a perpetual and many-sided play with our environment.

[18] W. Blair Bell, The Sex-Complex, 1920, p. 108. This book is a cautious and precise statement of the present state of knowledge on this subject, although some of the author's psychological deductions must be treated with circumspection.

It is thus that we arrive at the importance of the play-function, and thus, also, we realise that while it extends beyond the sexual sphere it yet definitely includes that sphere. There are at least three different ways of understanding the biological function of play. There is the conception of play, on which Groos has elaborately insisted, as education: the cat "plays" with the mouse and is thereby educating itself in the skill necessary to catch mice; all our human games are a training in qualities that are required in life, and that is why in England we continue to attribute to the Duke of Wellington the saying that "the battle of Waterloo was won on the playing fields of Eton." Then there is the conception of play as the utilisation in art of the superfluous energies left unemployed in the practical work of life;

this enlarging and harmonising function of play, while in the lower ranges it may be spent trivially, leads in the higher ranges to the production of the most magnificent human achievements. But there is yet a third conception of play, according to which it exerts a direct internal influence—health-giving, developmental, and balancing—on the whole organism of the player himself. This conception is related to the other two, and yet distinct, for it is not primarily a definite education in specific kinds of life-conserving skill, although it may involve the acquisition of such skill, and it is not concerned with the construction of objective works of art, although—by means of contact in human relationship—it attains the wholesome organic effects which may be indirectly achieved by artistic activities. It is in this sense that we are here concerned with what we may perhaps best call the play-function of sex.[19]

[19] The term seems to have been devised by Professor Maurice Parmelee, Personality and Conduct, 1918, pp. 104, 107, 113. But it is understood by Parmelee in a much vaguer and more extended sense than I have used it.

As thus understood, the play-function of sex is at once in an inseparable way both physical and psychic. It stimulates to wholesome activity all the complex and inter-related systems of the organism. At the same time it satisfies the most profound emotional impulses, controlling in harmonious poise the various mental instincts. Along these lines it necessarily tends in the end to go beyond its own sphere and to embrace and introduce into the sphere of sex the other two more objective fields of play, that of play as education, and that of play as artistic creation. It may not be true, as was said of old time, "most of our arts and sciences were invented for love's sake." But it is certainly true that, in proportion as we truly and wisely exercise the play-function of sex, we are at the same time training our personality on the erotic side and acquiring a mastery of the art of love.

The longer I live the more I realise the immense importance for the individual of the development through the play-function of erotic personality, and for human society of the acquirement of the art of love. At the same time I am ever more astonished at the rarity of erotic personality and the ignorance of the art of love even among those men and women, experienced in the exercise of procreation, in whom we might most confidently expect to find such development and such art. At times one feels hopeless at the thought that civilisation in this supremely intimate field of life has yet achieved so little. For until it is generally possible to acquire erotic personality and to master the art of loving, the development of the individual man or woman is marred, the acquirement of human happiness and harmony remains impossible.

In entering this field, indeed, we not only have to gain true knowledge but to cast off false knowledge, and, above all, to purify our hearts from superstitions which have no connection with any kind of existing knowledge. We have to cease to regard as admirable the man who regards the accomplishment of the procreative act, with the pleasurable relief it affords to himself, as the whole code of love. We have to treat with contempt the woman who abjectly accepts the act, and her own passivity therein, as the whole duty of love. We have to understand that the art of love has nothing to do with vice, and the acquirement of erotic personality nothing to do with sensuality. But we have also to realise that the art of love is far from being the attainment of a refined and luxurious self-indulgence, and the acquirement of erotic personality of little worth unless it fortifies and enlarges the whole personality in all its aspects. Now all this is difficult, and for some people even painful; to root up is a more serious matter than to sow; it cannot all be done in a day.

It is not easy to form a clear picture of the erotic life of the average man in our society. To the best informed among us knowledge in this field only comes slowly. Even when we have decided what may or may not be termed "average" the sources of approach to this intimate sphere remain few and misleading; at the best the women a man loves remain far more illuminating sources of information than the man himself. The more one knows about him, however, the more one is convinced that, quite independently of the place we may feel inclined to afford to him in the scale of virtue, his conception of erotic personality, his ideas on the art of love, if they have any existence at all, are of a humble character. As to the notion of play in the sphere of sex, even if he makes blundering attempts to practice it, that is for him something quite low down, something to be ashamed of, and he would not dream of associating it with anything he has been taught to regard as belonging to the spiritual sphere. The conception of "divine play" is meaningless to him. His fundamental ideas, his cherished ideals, in the erotic sphere, seem to be reducible to two: (1) He wishes to prove that he is "a man," and he experiences what seems to him the pride of virility in the successful attainment of that proof; (2) he finds in the same act the most satisfactory method of removing sexual tension and in the ensuing relief one of the chief pleasures of life. It cannot be said that either of these ideals is absolutely unsound; each is part of the truth; it is only as a complete statement of the truth that they become pathetically inadequate. It is to be noted that both of them are based solely on the physical act of sexual conjunction, and that they are both exclusively self-regarding. So that they are, after all, although the nearest approach to the erotic sphere he may be able to find, yet still not really erotic. For love is not primarily self-regarding. It is the intimate, harmonious, combined play—the play in the wide as well as in the more narrow sense we are here concerned with—of two personalities. It would not be love if it were primarily self-regarding, and the act of intercourse, however essential to secure the propagation of the race, is only an incident, and not an essential in love.

Let us turn to the average woman. Here the picture must usually be still more unsatisfactory. The man at least, crude as we may find his two fundamental notions to be, has at all events attained mental pride and physical satisfaction. The woman often attains neither, and since the man, by instinct or tradition, has maintained a self-regarding attitude, that is not surprising. The husband—by primitive instinct partly, certainly by ancient tradition—regards himself as the active partner in matters of love and his own pleasure as legitimately the prime motive for activity. His wife consequently falls into the complementary position, and regards herself as the passive partner and her pleasure as negligible, if not indeed as a thing to be rather ashamed of, should she by chance experience it. So that, while the husband is content with a mere simulacrum and pretence of the erotic life, the wife has often had none at all.

Few people realise—few indeed have the knowledge or the opportunity to realise—how much women thus lose, alike in the means to fulfill their own lives and in the power to help others. A woman has a husband, she has marital relationships, she has children, she has all the usual domestic troubles—it seems to the casual observer that she has everything that constitutes a fully developed matron fit to play her proper part in the home and in the world. Yet with all these experiences, which undoubtedly are an important part of life, she may yet remain on the emotional side—and, as a matter of fact, frequently remains—quite virginal, as immature as a school-girl. She has not acquired an erotic personality, she has not mastered the art of love, with the result that her whole nature remains ill-developed and unharmonised, and that she is incapable of bringing her personality—having indeed no achieved personality to bring—to bear effectively on the problems of society and the world around her.

That alone is a great misfortune, all the more tragic since under favourable conditions, which it should have been natural to attain, it might so easily be avoided. But there is this further result, full of the possibilities of domestic tragedy, that the wife so situated, however innocent, however virtuous, may at any time find her virginally sensitive emotional nature fertilised by the touch of some other man than her husband.

It happens so often. A girl who has been carefully guarded in the home, preserved from evil companions, preserved also from what her friends regarded as the contamination of sexual knowledge, a girl of high ideals, yet healthy and robust, is married to a man of whom she probably has little more than a conventional knowledge. Yet he may by good chance be the masculine counterpart of herself, well brought up, without sexual experience and ignorant of all but the elementary facts of sex, loyal and honourable, prepared to be, fitted to be, a devoted husband. The union seems to be of the happiest kind; no one detects that anything is lacking to this perfect marriage; in course of time one or more children are born. But during all this time the husband has never really made love to his wife; he has not even understood what courtship in the intimate sense means; love as an art has no existence for him; he has loved his wife according to his imperfect knowledge, but he has never so much as realised that his knowledge was imperfect. She on her side loves her husband; she comes in time indeed to have a sort of tender maternal feeling for him. Possibly she feels a little pleasure in intercourse with him. But she has never once been profoundly aroused, and she has never once been utterly satisfied. The deep fountains of her nature have never been unsealed; she has never been fertilised throughout her whole nature by their liberating influence; her erotic personality has never been developed. Then something happens. Perhaps the husband is called away, it may have been to take part in the Great War. The wife, whatever her tender solicitude for her absent partner, feels her solitude and is drawn nearer to friends, perhaps her husband's friends. Some man among them becomes congenial to her. There need be no conscious or overt love-making on either side, and if there were the wife's loyalty might be aroused and the friendship brought to an end. Love-making is not indeed necessary. The wife's latent erotic needs, while still remaining unconscious, have come nearer to the surface; now that she has grown mature and that they have been stimulated yet unsatisfied for so long, they have, unknown to herself, become insistent and sensitive to a sympathetic touch. The friends may indeed grow into lovers, and then some sort of solution, by divorce or intrigue—scarcely however a desirable kind of solution—becomes possible. But we are here taking the highest ground and assuming that honourable feeling, domestic affection, or a stern sense of moral duty, renders such solution unacceptable. In due course the husband returns, and then, to her utter dismay, the wife discovers, if she has not discovered it before, that during his absence, and for the first time in her life, she has fallen in love. She loyally confesses the situation to her husband, for whom her affection and attachment remain the same as before, for what has happened to her is the coming of a totally new kind of love and not any change in her old love. The situation which arises is one of torturing anxiety for all concerned, and it is not less so when all concerned are animated by noble and self-sacrificing impulses. The husband in his devotion to his wife may even be willing that her new impulses should be gratified. She, on her side, will not think of yielding to desires which seem both unfair to her husband and opposed to all her moral traditions. We are not here concerned to consider the most likely, or the most desirable, exit from this unfortunate situation. The points to note are that it is a situation which to-day actually occurs; that it causes acute unhappiness to at least two people who may be of the finest physical and intellectual type and the noblest character, and that it might be avoided if there were at the outset a proper understanding of the married state and of the part which the art of love plays in married happiness and the development of personality.

A woman may have been married once, she may have been married twice, she may have had children by both husbands, and yet it may not be until she is past the age of thirty and is united to a third man that she attains the development of erotic personality and all that it involves in the full flowering of her whole nature. Up to then she had to all appearance had all the essential experiences of life. Yet she had remained spiritually virginal, with conventionally prim ideas of life, narrow in her sympathies, with the finest and noblest functions of her soul helpless and bound, at heart unhappy even if not clearly realising that she was unhappy. Now she has become another person. The new liberated forces from within have not only enabled her to become sensitive to the rich complexities of intimate personal relationship, they have enlarged and harmonised her realisation of all relationships. Her new erotic experience has not only stimulated all her energies, but her new knowledge has quickened all her sympathies. She feels, at the same time, more mentally alert, and she finds that she is more alive than before to the influences of nature and of art. Moreover, as others observe, however they may explain it, a new beauty has come into her face, a new radiancy into her expression, a new force into all her activities. Such is the exquisite flowering of love which some of us who may penetrate beneath the surface of life are now and then privileged to see. The sad part of it is that we see it so seldom and then often so late.

It must not be supposed that there is any direct or speedy way of introducing into life a wider and deeper conception of the erotic play-function, and all that it means for the development of the individual, the enrichment of the marriage relationship, and the moral harmony of society. Such a supposition would merely be to vulgarise and to stultify the divine and elusive mystery. It is only slowly and indirectly that we can bring about the revolution which in this direction would renew life. We may prepare the way for it by undermining and destroying those degrading traditional conceptions which have persisted so long that they are instilled into us almost from birth, to work like a virus in the heart, and to become almost a disease of the soul. To make way for the true and beautiful revelation, we can at least seek to cast out those ancient growths, which may once have been true and beautiful, but now are false and poisonous. By casting out from us the conception of love as vile and unclean we shall purify the chambers of our hearts for the reception of love as something unspeakably holy.

In this matter we may learn a lesson from the psycho-analysts of to-day without any implication that psycho-analysis is necessarily a desirable or even possible way of attaining the revelation of love. The wiser psycho-analysts insist that the process of liberating the individual from outer and inner influences that repress or deform his energies and impulses is effected by removing the inhibitions on the free-play of his nature. It is a process of education in the true sense, not of the suppression of natural impulses nor even of the instillation of sound rules and maxims for their control, not of the pressing in but of the leading out of the individual's special tendencies.[20] It removes inhibitions, even inhibitions that were placed upon the individual, or that he consciously or unconsciously placed upon himself, with the best moral intentions, and by so doing it allows a larger and freer and more natively spontaneous morality to come into play. It has this influence above all in the sphere of sex, where such inhibitions have been most powerfully laid on the native impulses, where the natural tendencies have been most surrounded by taboos and terrors, most tinged with artificial stains of impurity and degradation derived from alien and antiquated traditions. Thus the therapeutical experience of the psycho-analysts reinforces the lessons we learn from physiology and psychology and the intimate experiences of life.

[20] See, for instance, H.W. Frink, Morbid Fears and Compulsions, 1918, Ch. X.

Sexual activity, we see, is not merely a bald propagative act, nor, when propagation is put aside, is it merely the relief of distended vessels. It is something more even than the foundation of great social institutions. It is the function by which all the finer activities of the organism, physical and psychic, may be developed and satisfied. Nothing, it has been said, is so serious as lust—to use the beautiful term which has been degraded into the expression of the lowest forms of sensual pleasure—and we have now to add that nothing is so full of play as love. Play is primarily the instinctive work of the brain, but it is brain activity united in the subtlest way to bodily activity. In the play-function of sex two forms of activity, physical and psychic, are most exquisitely and variously and harmoniously blended. We here understand best how it is that the brain organs and the sexual organs are, from the physiological standpoint, of equal importance and equal dignity. Thus the adrenal glands, among the most influential of all the ductless glands, are specially and intimately associated alike with the brain and the sex organs. As we rise in the animal series, brain and adrenal glands march side by side in developmental increase of size, and at the same time, sexual activity and adrenal activity equally correspond.

Lovers in their play—when they have been liberated from the traditions which bound them to the trivial or the gross conception of play in love—are thus moving amongst the highest human activities, alike of the body and of the soul. They are passing to each other the sacramental chalice of that wine which imparts the deepest joy that men and women can know. They are subtly weaving the invisible cords that bind husband and wife together more truly and more firmly than the priest of any church. And if in the end—as may or may not be—they attain the climax of free and complete union, then their human play has become one with that divine play of creation in which old poets fabled that, out of the dust of the ground and in his own image, some God of Chaos once created Man.

THE INDIVIDUAL AND THE RACE

The relation of the individual person to the species he belongs to is the most intimate of all relations. It is a relation which almost amounts to identity. Yet it somehow seems so vague, so abstract, as scarcely to concern us at all. It is only lately indeed that there has been formulated even so much as a science to discuss this relationship, and the duties which, when properly understood, it throws upon the individual. Even yet the word "Eugenics," the name of this science, and this art, sometimes arouses a smile. It seems to stand for a modern fad, which the superior person, or even the ordinary plebeian democrat, may pass by on the other side with his nose raised towards the sky. Modern the science and art of Eugenics certainly seem, though the term is ancient, and the Greeks of classic days, as well as their successors to-day, used the word Eugeneia for nobility or good birth. It was chosen by Francis Galton, less than fifty years ago, to express "the effort of Man to improve his own breed." But the thing the term stands for is, in reality, also far from modern. It is indeed ancient and may even be nearly as old as Man himself. Consciously or unconsciously, sometimes under pretexts that have disguised his motives even from himself, Man has always been attempting to improve his own quality or at least to maintain it. When he slackens that effort, when he allows his attention to be too exclusively drawn to other ends, he suffers, he becomes decadent, he even tends to die out.

Primitive eugenics had seldom anything to do with what we call "birth-control." One must not say that it never had. Even the mysterious mika operation of so primitive a race as the Australians has been supposed to be a method of controlling conception. But the usual method, even of people highly advanced in culture, has been simpler. They preferred to see the new-born infant before deciding whether it was likely to prove a credit to its parents or to the human race generally, and if it seemed not up to the standard they dealt with it accordingly. At one time that was regarded as a cruel and even inhuman method. To-day, when the most civilised nations of the world have devoted all their best energies to competitive slaughter, we may have learnt to view the matter differently. If we can tolerate the wholesale murder and mutilation of the finest specimens of our race in the adult possession of all their aptitudes we cannot easily find anything to disapprove in the merciful disposal of the poorest specimens before they have even attained conscious possession of their senses. But in any case, and whatever we may ourselves be pleased to think or not to think, it is certain that some of the most highly developed peoples of the world have practised infanticide. It is equally certain that the practise has not proved destructive to the emotions of humanity and affection. Even some of the lowest human races,—as we commonly estimate them,—while finding it necessary to put aside a certain proportion of their new-born infants, expend a degree of love and even indulgence on the children they bring up which is rarely found among so-called civilised nations.

There is no need, however, to consider whether or not infanticide is humane. We are all agreed that it is altogether unnecessary, and that it is seldom that even that incipient form of infanticide called abortion, still so popular among us, need be resorted to. Our aim now—so far at all events as mere ideals go—is not to destroy life but to preserve it; we seek to improve the conditions of life and to render unnecessary the premature death of any human creature that has once drawn breath.

It is indeed just here that we find a certain clash between the modern view of life and the view of earlier civilisations. The ancients were less careful than we claim to be of the individual, but they were more careful of the race. They cultivated eugenics after their manner, though it was a manner which we reprobate.[21] We pride ourselves, rightly or wrongly, on our care for the individual; during all the past century we claim to have been strenuously working for an amelioration of the environment which will make life healthier and pleasanter for the individual. But in the concentration of our attention on this altogether desirable end, which we are still far from having adequately attained, we have lost sight of that larger end, the well-being of the race and the amelioration of life itself, not merely of the conditions of life. The most we hope is that somehow the improvement of the conditions of the individual will incidentally improve the stock. These our practical ideals, which have flourished for a century past, arose out of the great French Revolution and were inspired by the maxim of that Revolution, as formulated by Rousseau, that "All men are born equal." That maxim, was overthrown half a century ago; the great biological movement of science, initiated by Darwin, showed that it was untenable. All men are not born equal. Everyone agrees about that now, but nevertheless the momentum of the earlier movement was so powerful that we still go on acting as though all men are, and always will be, born equal, and that we need not trouble ourselves about heredity but only about the environment.

[21] But this statement must not be left without important qualification. Thus the ancient Greeks (as Moïssidès has shown in Janus, 1913), not only their philosophers and statesmen, but also their women, often took the most enlightened interest in eugenics, and, moreover, showed it in practice. They were in many respects far in advance of us. They clearly realised, for instance, the need of a proper interval between conceptions, not only to ensure the health of women, but also the vigour of the offspring. It is natural that among every fine race eugenics should be almost an instinct or they would cease to be a fine race. It is equally natural that among our modern degenerates eugenics is an unspeakable horror, however much, as the psycho-analysts would put it, they rationalise that horror.

The way out of this clash of ideals—which has compelled us to hope impossibilities from the environment because we dreaded what seemed the only alternative—is, as we know, furnished by birth-control. An unqualified reliance on the environment, making it ever easier and easier for the feeblest and most defective to be born and survive, could only, in the long run, lead to the degeneration of the whole race. The knowledge of the practice of birth-control gives us the mastery of all that the ancients gained by infanticide, while yet enabling us to cherish that ideal of the sacredness of human life which we profess to honour so highly. The main difficulty is that it demands a degree of scientific precision which the ancients could not possess and might dispense with, so long as they were able to decide the eugenic claims of the infant by actual inspection. We have to be content to determine not what the infant is but when it be likely to be, and that involves a knowledge of the laws of heredity which we are only learning slowly to acquire. We may all in our humble ways help to increase that knowledge by giving it greater extension and more precision through the observations we are able to make on our own families. To such observations Galton attached great importance and strove in various ways to further them. Detailed records, physical and mental, beginning from birth, are still far from being as common as is desirable, although it is obvious that they possess a permanent personal and family private interest in addition to their more public scientific value. We do not need, and it would indeed be undesirable, to emulate in human breeding the achievements of a Luther Burbank. We have no right to attempt to impose on any human creature an exaggerated and one-sided development. But it is not only our right, it is our duty, or rather one may say, the natural impulse of every rational and humane person, to seek that only such children may be born as will be able to go through life with a reasonable prospect that they will not be heavily handicapped by inborn defect or special liability to some incapacitating disease. What is called "positive" eugenics—the attempt, that is, to breed special qualities—may well be viewed with hesitation. But so-called "negative" eugenics—the effort to clear all inborn obstacles out of the path of the coming generation—demands our heartiest sympathy and our best co-operation, for as Galton, the founder of modern Eugenics, wrote towards the end of his life of this new science: "Its first object is to check the birth-rate of the unfit, instead of allowing them to come into being, though doomed in large numbers to perish prematurely." We can seldom be absolutely sure what stocks should not propagate, and what two stocks should on no account be blended, but we can attain reasonable probability, and it is on such probabilities in every department of life that we are always called upon to act.

It is often said—I have said it myself—that birth-control when practised merely as a limitation of the family, scarcely suffices to further the eugenic progress of the race. If it is not deliberately directed towards the elimination of the worst stocks or the worst possibilities in the blending of stocks, it may even tend to diminish the better stocks since it is the better stocks that are least likely to propagate at random. This is true if other conditions remain equal. It is evident, however, that the other conditions will not remain equal, for no evidence has yet been brought forward to show that birth-control, even when practised without regard to eugenic considerations—doubtless the usual rule up to the present—has produced any degeneration of the race. On the contrary, the evidence seems to show that it has improved the race. The example of Holland is often brought forward as evidence in favour of such a tendency of birth-control, since in that country the wide-spread practise of birth-control has been accompanied by an increase in the health and stature of the people, as well as an increase in their numbers to a remarkable degree, for the fall in the birth-rate has been far more than compensated by the fall in the death-rate, while it is said that the average height of the population has increased by four inches. It is, indeed, quite possible to see why, although theoretically a random application of birth-control cannot affect the germinal possibilities of a community, in practise it may improve the somatic conditions under which the germinal elements develop. There will probably be a longer interval between the births of the children, which has been demonstrated by Ewart and others to be an important factor not only in preserving the health of the mother but in increasing the health and size of the child. The diminution in the number of the children renders it possible to bestow a greater amount of care on each child. Moreover, the better economic position of the father, due to the smaller number of individuals he has to support, makes it possible for the family to live under improved conditions as regards nourishment, hygiene, and comfort. The observance of birth-control is thus a far more effective lever for raising the state of the social environment and improving the conditions of breeding, than is direct action on the part of the community in its collective capacity to attain the same end. For however energetic such collective action may be in striving to improve general social conditions by municipalising or State-supporting public utilities, it can never adequately counter-balance the excessive burden and wasteful expenditure of force placed on a family by undue child-production. It can only palliate them.

When, however, we have found reason to believe that, even if practised without regard to eugenic considerations, birth-control may yet act beneficially to promote good breeding, we begin to realise how great a power it may possess when consciously and deliberately directed towards that end. In eugenics, as already pointed out, there are two objects that may be aimed at: one called positive eugenics, that seeks to promote the increase of the best stocks amongst us; the other, called negative eugenics, which seeks to promote the decrease of the worst stocks. Our knowledge is still too imperfect to enable us to pursue either of these objects with complete certainty. This is especially so as regards positive eugenics, and since it seems highly undesirable to attempt to breed human beings, as we do animals, for points, when we are in the presence of what seem to us our finest human stocks, physically, morally, and intellectually, it is our wisest course just to leave them alone as much as we can. The best stocks will probably be also those best able to help themselves and in so doing to help others. But that is obviously not so as regards the worst stocks. It is, therefore, fortunate that the aim here seems a little clearer. There are still many abnormal conditions of which we cannot say positively that they are injurious to the race and that we should therefore seek to breed them out. But there are other conditions so obviously of evil import alike to the subjects themselves and to their descendants that we cannot have any reasonable doubt about them. There is, for instance, epilepsy, which is known to be transformed by heredity into various abnormalities dangerous alike to their possessors and to society. There are also the pronounced degrees of feeble-mindedness, which are definitely heritable and not only condemn those who reveal them to a permanent inaptitude for full life, but constitute a subtle poison working through the social atmosphere in all directions and lowering the level of civilisation in the community. Nowhere has this been so thoroughly studied and so clearly proved as in the United States. It is only necessary to mention Dr. C.B. Davenport of the Department of Experimental Evolution at Cold Spring Harbor (New York) who has carried on so much research in regard to the heredity of epilepsy and other inheritable abnormal conditions, and Dr. Goddard of Vineland (New Jersey) whose work has illustrated so fully the hereditary relationships of feeble-mindedness. The United States, moreover, has seen the development of the system of social field-work which has rendered possible a more complete knowledge of family heredity than has ever before been possible on a large scale.

It is along such lines as these that our knowledge of the eugenic conditions of life will grow adequate and precise enough to form an effective guide to social conduct. Nature, and a due attention to laws of heredity in life, will then rank in equal honour to our eyes with nurture or that attention to the environmental conditions of life which we already regard as so important. A regard to nurture has led us to spend the greatest care on the preservation not only of the fit but the unfit, while meantime it has wisely suggested to us the desirability of segregating or even of sterilising the unfit. But the study of Nature leads us further and, as Galton said, "Eugenics rests on bringing no more individuals into the world than can be properly cared for, and these only of the best stocks." That is to say that the only instrument by which eugenics can be made practically effective in the modern world is birth-control.

It is not scientific research alone, nor even the wide popular diffusion of knowledge, that will suffice to bring eugenics and birth-control, singly or in their due combination, into the course of our daily lives. They need to be embodied in our instinctive impulses. Galton considered that eugenics must become a factor of religion and be regarded as a sacred and virile creed, while Ellen Key holds that the religions of the past must be superseded by a new religion which will be the awakening of the whole of humanity to a consciousness of the "holiness of generation." For my own part, I scarcely consider that either eugenics or birth-control can be regarded as properly a part of religion. Being of virtue and not of grace they belong more naturally to the sphere of morals. But here they certainly need to go far deeper than the mere intelligence of the mind can take them. They cannot become guides to conduct until their injunctions have been printed on the fleshy tablets of our hearts. The demands of the race must speak from within us, in the voice of conscience which we disobey at our peril. When that happens with regard to ascertained laws of racial well-being we may know that we are truly following, even though not in the letter, those great spirits, like Galton with his intellectual vision and Ellen Key with her inspired enthusiasm, who have pointed out new roads for the ennoblement of the race.

It may be well, before we go further, to look a little more closely into the suspicion and dislike which eugenics still arouses in many worthy old-fashioned people. To some extent that attitude is excused, not only by the mistakes which in a new and complex science must inevitably be made even by painstaking students, but also by the rash and extravagant proposals of irresponsible and eccentric persons claiming without warrant to speak in the name of eugenics. Two thousand years ago the wild excesses of some early

Christians furnished an excuse for the ancient world to view Christianity with contempt, although the extreme absence of such excesses has furnished still better ground for the modern world to maintain the same view. To-day such a work as Le Haras Humain ("The Human Stud-farm") of Dr. Binet-Sanglé, putting forward proposals which, whether beneficial or not, will certainly find no one to carry them out, similarly furnishes an excuse to those who would reject eugenics altogether. Utopian schemes have their value; we should be able to find inspiration in the most modern of them, just as we still do in Plato's immortal Republic. But in this, as in other matters, we must exercise a little intelligence. We must not confuse the brilliant excursion of some solitary thinker with the well-grounded proposals of those who are concerned with the sober possibilities of actual life in our own time. People who are incapable of exercising a little shrewd commonsense in the affairs of life, and are in the habit of emptying out the baby with the bath, had better avoid touching the delicate problems connected with practical eugenics.

There is one prejudice already mentioned, due to lack of clear thinking, which deserves more special consideration because it is widespread among the socialistic democracy of several countries as well as among social reformers, and is directed alike against eugenics and birth-control. This prejudice is based on the ground that bad economic conditions and an unwholesome environment are the source of all social evils, and that a better distribution of wealth, or a vast scheme of social welfare, is the one thing necessary, when that is achieved all other things being added unto us, without any further trouble on our part. It is certainly impossible to over-rate the importance of the economic factor in society, or of a good environment. And it is true that eugenics alone, like birth-control alone, can effect little if the economic basis of society is unsound. But it is equally certain that the economic factor can never in itself suffice for fine living or even as a cure-all of social and racial diseases. Its value is not that it can effect these things but that it furnishes the favourable conditions for effecting them. He would be foolish indeed who went to the rich to find the example of good breeding and, as is well known, it is not with the rich that the future of the race lies. The fact is that under any economic system the responsible personal direction of the individual and the family remain equally necessary, and no progress is possible so long as the individual casts all responsibility away from himself on to the social group he forms part of. The social group, after all, is merely himself and the likes of himself. He is merely shifting the burden from his individual self to his collective self, and in so doing he loses more than he gains.

Thus there is always a sound core in that Individualism which has been preached so long and practised so energetically, especially in English-speaking lands, however great the abuse involved in its excesses. It is still in the name of Individualism that the most brilliant antagonists of eugenics and of birth-control are wont to direct their attacks. The counsel of self-control and foresight in procreation, the restriction necessary to purify and raise the standard of the race, seem to the narrow and short-sighted advocates of a great principle an unwarrantable violation of the sacred rights of their individual liberty. They have not yet grasped the elementary fact that the rights of the individual are the rights of all individuals, and that Individualism itself calls for a limitation of the freedom of the individual.

That is why even the most uncompromising Individualist must recognise an element of altruism, call it whatever name you will, Collectivism, Socialism, Communism, or merely the vague and long-suffering term, Democracy. One cannot assume Individualism for oneself unless one assumes it for the many. That is a great truth which goes to the heart of the whole complex problem of eugenics and birth-control. As Perrycoste has well argued,[22] biology is altogether against the narrow Individualism which seeks to oppose Collective Individualism. For if, in accordance with the most careful modern investigations, we recognise that heredity is supreme, that the qualities we have inherited from our ancestors count for more in our lives than anything we have acquired by our own personal efforts, then we have to admit that the capable man's wealth is more the community's property than his own, and, similarly, the incapable man's poverty is more the community's concern than his own. So that neither the capable nor the incapable are entitled to an unqualified power of freedom, and neither, likewise, are justly liable to be burdened by an unqualified responsibility. It is the duty of the community to draw on the powers of the fit and equally its duty to care for the unfit. In this way, Perrycoste, whose attitude is that of the Rationalist, is led by science to a conclusion which is that of the Christian. We are all members each of the other, and still more are we members of those who went before us. The generations preceding us have not died to themselves but live in us, and we, whom they produced, live in each other and in those who will come after us. The problems of eugenics and of birth-control affect us all. In the face of these problems it is the voice of Man that speaks: "Inasmuch as ye did it not unto the least of these my brethren, ye did it not unto me." However firmly we base ourselves on the principles of Individualism we are inevitably brought to the fundamental facts of eugenics which, if we fail to recognise, our Individualism becomes of no effect.

[22] F.H. Perrycoste, "Politics and Science," Science Progress, Jan., 1920.

But it is the same with Socialism, or by whatever name we chose to call the Collectivist activities of the community in social reform. Socialism also brings us up against the hard rock of eugenic fact which, if we neglect it, will dash our most beautiful social construction to fragments. It is the more necessary to point this out since it is on the Socialist and Democratic side, much more frequently than on the Individualist side, that we find an indifferent or positively hostile attitude towards eugenic considerations. Put social conditions on a sound basis, the people on this side often say, let all receive an adequate economic return for their work and be recognised as having a claim for an adequate share in the products of society, and there is no need to worry about the race or about the need for birth-control, all will go well of itself. There is not the slightest ground for any such comfortable belief.

This has been well shown by Dr. Eden Paul, himself a Socialist and even in sympathy with the extreme Left.[23] After setting forth the present conditions, with our excessive elimination of higher types, and undue multiplication of lower types, the racial degeneration caused by the faulty and anti-selective working of the marriage system in modern capitalist society, so that in our existing civilisation unconscious natural selection has largely ceased to work towards the improvement of the human breed, he proceeds to consider the possible remedies. The frequent impatience of the Socialist, and Social Reformers generally, with eugenic proposals has a certain degree of justification in the fact that many evils thoughtlessly attributed to inferiority of stock are really due to bad environment. But when the environment has been so far improved that all defects due to its badness are removed, we shall be face to face, without possibility of doubt, with bad inheritance as the sole remaining factor in the production of inefficient and anti-social members of the community. A socialist community must recognise the right to work and to maintenance of all its members, Eden Paul points out, but, he adds, a community which allowed this right to all defectives without imposing any restrictions in their perpetuation of themselves would deserve all the evils that would fall upon it. It is quite clear how intolerable the burden of these evils would be. A State that provided an adequate subsistence for all alike, the inefficient as well as the efficient, would encourage a racial degeneration, from excessive multiplication of the unfit, far more dangerous even than that of to-day.[24] Ability to earn the minimum wage, Eden Paul argues in agreement with H.G. Wells, must be the condition of the right to become a parent. "Unless the socialist is a eugenist as well, the socialist state will speedily perish from racial degradation."

[23] In an essay on "Eugenics, Birth Control, and Socialism" in Population and Birth-Control: A Symposium, edited by Eden and Cedar Paul.

[24] This is here and there beginning to be recognised. Thus, not long ago, the Hereford War Pensions Committee resolved not to issue a maternal grant for children born during a prolonged period of treatment allowance. Such a measure of course fails to meet the situation, for it is obvious that, when born, the children must be cared for. But it shows a glimmering recognition of the facts, and the people capable of such a recognition will,

in time, come to see that the right way of meeting the situation is, not to neglect the children, but to prevent their conception. Mothers' Clinics for instruction in such prevention are now being established in England, through the advocacy of Mrs. Margaret Sanger and the actual initiative of Dr. Marie Stopes.

Thus it is essential that the eugenist, dealing with the hereditary factor of life, and the social reformer or socialist, dealing with the environmental factor, should supplement each other's work. Neither can attain his end without the other's help, for the eugenist alone cannot overcome the environmental factor, even perhaps increases it if he is an individualist in the narrow sense, and the socialist alone cannot overcome the bad hereditary factor, and will even increase it if he is no more than a socialist. The more socialist our State becomes the more essential becomes at the same time the adoption of eugenic practices as a working part of the State. "Socialism and eugenics must go hand in hand."

Perrycoste from his own point of view has independently reached the same conclusions. He is not, indeed, concerned with any "Socialist" community of the future but with the dangerous results which must inevitably follow the already established methods of social reform in our modern civilised States unless they are speedily checked by effective action based on eugenic knowledge. "If," he observes, "the community is to shoulder half or three-quarters of the burden of sustaining those degenerates who, through no fault of their own, are congenitally incompetent to maintain themselves in decent comfort, and is to render the life-pilgrimage of these unfortunates tolerable instead of a dreary nightmare, if it is to assume paternal charge of all the tens or hundreds of thousands of children whose parents cannot or will not provide adequately for them and is to guarantee to all such children as much education as they are capable of receiving, and a really fair start in life: then in sheer self-preservation the community must insist on, and rigidly enforce, its absolute claim to secure that no degeneracy or inheritable congenital defects shall persist beyond the present generation of degenerates, and that the community of fifty or seventy years hence shall have no incubus of mentally, or morally, or even physically, degenerate members—none but a few occasional sporadic morbid 'sports' from the normal, which it, in turn, may effectively prevent from handing on their like." Unless the problem is squarely faced, Perrycoste concludes, national deterioration must increase and a permanently successful collectivist society is inherently impossible.

We are not now concerned with the details of any policy of eugenics and of birth-control, which I couple together because although a random birth-control by no means involves much, if any, eugenic progress, it is not easy under modern conditions to conceive any practical or effective policy of eugenics except through the instrumentation of birth-control. We here take it for granted that in this field the slow progress of scientific knowledge must be our guide. Premature legislation, rash and uninstructed action, will not lead to progress but are more likely to delay it. Yet even with imperfect knowledge, it is already of the first importance to evoke interest in the great issue here at stake and to do all that we can to arouse the individual conscience of every man and woman to his or her personal responsibility in this matter. That is here all taken for granted.

It seems necessary to consider the political aspect of eugenics because that aspect is frequently invoked, and a man's attitude towards this question is frequently determined beforehand by what he considers that Individualism or Socialism demands. We see that when the question is driven home our political attitude makes no difference. It is only a shallow Individualism, it is only a still more shallow Socialism, which imagines that under modern social conditions the fundamental racial questions can be left to answer themselves.

Many years before the Great War, in all the most civilised countries of the World, there were those who raised the cry of "Race-Suicide!" In America this cry was more especially popularised by the powerful voice of Theodore Roosevelt, but in European countries there were similar voices raised in tones of virtuous indignation to denounce the same crime. Since the war other voices have been raised in even more high-pitched and feverish tones, but now they are less weighty and responsible voices, since to those who realise that at present there is not food enough to keep the population of the world from starvation it seems hardly compatible with sanity to advocate an increased rate of human production.

Now, though it is easy to do so, we must not belittle this cry of "Race-Suicide!" It is not usually accompanied by definite argument, but it assumes that birth-control is the method of such suicide, and that the first and most immediately dangerous result is that one's own nation, whichever that may be, is placed in a position of alarming military inferiority to other nations, as a step towards the final extinction. It is useless to deny that it really is a serious matter if there is danger of the speedy disappearance of the human race from the earth by its own voluntary and deliberate action, and that within a measurable period of time—for if it were an immeasurable period there would be no occasion for any acute anxiety—the last man will perish from the world. This is what "Race-Suicide" means, and we must face the fact squarely.

It can scarcely be said, however, that the meaning of "Race-Suicide" has actually been squarely faced by those who have most vehemently raised that cry. Translated into more definite and precise terms this cry means, and is intended to mean: "We want more births." That is what it definitely means, and sometimes in the minds of those who make this demand it seems also to imply nothing more. Yet it implies a great number of other things. It implies certain strain and probable ill-health on the mothers, it implies distress and disorder in the family, it implies, even if the additional child survives, a more acute industrial struggle, and it further involves in this case, by the stimulus it gives to over-population, the perpetual menace of militarism and war. What, however, even at the outset, more births most distinctly and most unquestionably imply is more deaths. It is nowadays so well known that a high birth-rate is accompanied by a high death-rate—the exceptions are too few to need attention—that it is unnecessary to adduce further evidence. It is only the intoxicated enthusiasts of the "Race-Suicide" cry who are able to overlook a fact of which they can hardly be ignorant. The model which they hold up for the public's inspiration has on the obverse "More Births!" But on the reverse it bears "More Deaths!" It would be helpful to the public, and might even be wholesome for our enthusiasts' own enlightenment, if they would occasionally turn the medal round and slightly vary the monotony of their propaganda by changing its form and crying out for "More Deaths!" "It is a hard thing," said Johnny Dunn, "for a man that has a house full of children to be left to the mercy of Almighty God."

If, however, we wish to consider the real significance of the facts, without regard for the wild cries of ignorant cranks, it is scarcely necessary to point out here that neither the birth-rate taken by itself, nor the death-rate taken by itself, will suffice to give us any measure even of the growth of the population, to say nothing of the progress of civilisation or the happiness of humanity. It is obvious that we must consider both gains and losses, and put one against the other, if we wish to ascertain the net result. We may roughly get a notion of what that result is by deducting the death-rate from the birth-rate and calling the remainder the survival-rate. If we are really concerned with the question of the alleged suicide of the race, and do not wish to be befooled, we must pay little attention to the birth-rate, for that by itself means nothing: we must concentrate on the survival-rate. Then we may soon convince ourselves, not only that the human race is not committing suicide, but that not even a single one of the so-called civilised nations of which it is mainly composed is committing suicide. Quite the contrary! Every one of them, even France, where this peculiar "suicide" is supposed to be most actively at work, is yearly increasing in numbers.

It is interesting to note, moreover, that the French have been increasing faster, that is to say the survival-rate has been higher in recent years just before the war, when the birth-rate was at its lowest, than they were twenty years earlier, with a higher birth-rate. And if we take a wider sweep and consider the growth of the French population towards the end of the eighteenth century, we find the birth-rate estimated at the very high figure of 40. But the death-rate was nearly as high, the average duration of life was only half what it is now. So that the survival-rate in France at that time, with widely different rates of birth and death, was not much unlike it is now. The recent French birth-rate of 19 and less, which automatically causes the "Race-Suicide" marionette to dance with rage, is producing not far from the same result in growth of the population—we are not here concerned with the enormous difference in well being and happiness—as the extremely high rate of 40 which sends our marionettes leaping to the sky with joy. In war-time England, in 1917, the birth-rate sank to 17.8, yet the death-rate was at 14 and the increase of the population continued. The more the human race commits this kind of suicide, one is tempted to exclaim, the faster it grows!

It is, however, in the New World—as in Canada, Australia, and New Zealand—that we find the most impressive evidence of the real criteria of the growth in population set up for judgment on the racial suicide cranks. Canadian statistics bring out many points instructive even in their variation. Here we see not only unusual curves of rise and fall, but also pronounced differences, due to the special peculiarities of the French population, most clearly in the Province of Quebec but also in some parts of the Province of Ontario. In Quebec the birth-rate some years ago was 35, and the death-rate 21, both rates high, and the survival-rate high at 14; recently the birth-rate has risen to 37 and the death-rate fallen to 17, with the result that the survival-rate of 20 is the highest in the world, though it must be noted that the high birth-rate is not likely to last long, since in Quebec, as elsewhere in the world, increasing urbanisation causes a decreasing birth-rate. In mainly English-speaking Ontario the birth-rate is much lower, about 24, but the death-rate is also lower, about 14, so that the fairly considerable survival-rate of 10 is obtained. But we note the highly significant fact that some thirty years or more ago the birth-rate was much lower, about 19, and yet the survival-rate was almost 9, nearly as high as to-day! The death-rate was then at 10, and nothing could be more instructive as to the real relationship that holds in this matter. There has been a great rise in the birth-rate and the only result, as someone has remarked, is a great increase in the population of the grave-yards. Equally instructive is it to compare various cities in this same Province, living under the same laws, and fairly similar social conditions. In the report of the Registrar-General of Ontario for 1916 I find that highest in birth-rate of cities in the Province stands Ottawa with a very considerable French population. But first also stands the same city for infant mortality, which is three times greater than in some other cities in the Province with a low birth-rate. Sault Ste. Marie, again with an enormous birth-rate, stands third for infant mortality. Canada shows us that, even if we regard the crude desire for a large growth of population as reasonable—and that is a considerable assumption—a high birth-rate is an uncertain prop to rest on.

Canada is an instructive example because we have some ground for believing that the difference between the English-speaking and French-speaking populations—the greater care of the former in procreation and the more recklessly destructive methods of the latter in attaining the same ends—are due to their different attitudes towards the use of methods of birth-control. What the result of a general use of such methods is we know from the example already mentioned of Holland, where they are taught, officially recognised, and in general use, not only among the rich but among the poor. The result is that the birth-rate has been falling slowly and steadily for forty years. But the death-rate has also been falling and at a greater rate. So that the more the birth-rate has fallen the higher has been the rate of increase among the population.

It is perhaps in Australia and New Zealand that we find the most satisfactory proofs of the benefits of a falling birth-rate in relation to "Race-Suicide." The evidence may well appeal to us the more since it is precisely here that the race-suicide fanatic finds freest scope for his wrath. He looks gleefully at China with its prolific women, at Russia with its magnificent birth-rate before the War of nearly 50, at Roumania with its birth-rate of 42, at Chile and Jamaica with nearly 40. No nonsense about birth-control there! No shirking by women of the sacred duties of perpetual maternity! No immoral notions about claims to happiness and desires for culture. And then he turns from, those great centres of prosperity and civilisation to Australia, to New Zealand, and his voice is choked and tears fill his eyes as he sees the goal of "Race-Suicide" nearly in sight and the spectre of the Last Man rising before him. For there is no doubt about it, Australia and New Zealand contain a population which is gradually reaching the highest point yet known of democratic organisation and general social well-being, and the birth-rate has been falling with terrific speed. Sixty-years ago in the Australian Commonwealth it was nearly 44, only forty years ago in New Zealand it was 42. Now it is only about 26 in both lands. Yet the survival-rate, the actual growth of the population, is not so very much less with this low birth-rate than it was with the high birth-rate. For the death-rate has also fallen in both lands to about 10 (in New Zealand to 9) which is lower than any other country in the world. The result is that Australia and New Zealand, where (so it is claimed) preventives of conception are hawked from door to door, instead of being awful examples of "Race-Suicide," actually present the highest rate of race-increase in the world (only excepting Canada, where it is less firmly and less healthily based), nearly twice that of Great Britain and able at the present rate to double itself every 44 years. So much for "Race-Suicide."

The outcry about "Race-Suicide" is so far away from the real facts of life that it is not easy to take it seriously, however solemn one's natural temperament may be. We are concerned with people who arrogantly claim to direct the moral affairs of the world, even in the most intimately private matters, and who are yet ignorant of the most elementary facts of the world, unable to think, not even able to count! We can only greet them with a smile. But this question has, nevertheless, a genuinely serious aspect, and I should be sorry even to touch on the question of birth-control in relation to "Race-Suicide" without making that serious aspect clear.

"Race-Suicide," we know, has no existence. Not only is the race as a whole increasing in number, especially its White branches, but even among the separate national groups there is not even one civilised people anywhere in the world that is decreasing in number. On the contrary they are all, even France, increasing at a more or less rapid rate. In England and Wales, for example, where the birth-rate has steadily fallen during the last forty years from 36 to 23 (I disregard the abnormal rates of War-time) the population is still increasing, and even if the present falls in birth-rate and death-rate continue, it will for years still go on increasing by an excess of over 1,000 births a day. When we realise that this is merely what goes on in one corner of the world and must be multiplied enormously to represent the whole, we shall find it impossible even to conceive the prodigious flow of excess babies which is being constantly poured over the earth. If we are capable of realising all the problems which thereby arise we must be forced to ask ourselves: Is this state of things desirable?

"Be ye fruitful and multiply." That command was, according to the old story, delivered to a world inhabited by eight people. It has been handed down to a world in which it has long been ridiculously out of place, and has become merely the excuse for criminal recklessness among a race which has chosen to forget that the command was qualified by a solemn admonition: "At the hand of man, even at the hand of every man's brother, will I require the life of man." The high birth-rate has meant a vast slaughter of infants, it has meant, moreover, a perpetual oppression of the workers, disease, starvation, and death among the adult population; it has meant, further, a blood-thirsty economic competition, militarism, warfare. It has meant that all civilisation has from time to time become a thin crust over a volcano of revolution, and the human race has gone on lightly dancing there, striving to forget that ancient warning from a soul of things even deeper than the voice of Jehovah: "At the hand of man will I require the life of man." Men have recklessly followed the Will o' the Wisp which represented mere multiplication of their inefficient selves as the ideal of progress, quantity before quality, the notion that in an orgy of universal procreation could consist the highest good of humanity.

The Great War, that is scarcely yet merged into an only less war-like Peace, has brought at least the small compensation that it has led men to look in the face this insane ideal of human progress. We see to-day what has come of it, and the further evils yet to come of it are being embodied beneath our eyes. So that at last the voice of Jehovah has here and there been faintly heard, even where nowadays we had grown least accustomed to hear it, in the Churches. It is Dr. Inge, the Dean of London's Cathedral of St. Paul's, a distinguished Churchman and at the same time a foremost champion of eugenics, who lately expressed the hope that the world, especially the European world, would one day realise the advantages of a stationary population.[25] Such a recognition, such an aspiration, indicates that a new hope is dawning on the world's horizon, and a higher ideal growing within the human soul. The mad competition of the industrial world during the past century, with the sordid gloom and wretchedness of it for all who were able to see beneath the surface, has shown for ever what comes of the effort to produce a growing population by high birth-rates in peace-time. The Great War of a later day has shown, let us hope in an equally decisive manner, what comes to a world where men have been for long generations produced so copiously and so cheaply that it is natural to regard them as only fit to sweep off the earth with machine guns. And the whole world of to-day—with its starving millions struggling in vain to feed themselves, with most of its natural beauty swept away by the ravages of man, and many of its most exquisite animals finally exterminated—is likely to become merely the monument to an ideal that failed. It was time, however late in the day, for a return to common-sense. It was time to realise that the ideal of mere propagation could lead us nowhere but to destruction. On that level we cannot compete even with the lowest of organised things, not even with the bacteria, which in number and in rapidity of multiplication are inconceivable to us. "All hope abandon, ye that enter here" is written over the portal of this path of "Progress."

[25] This has long been recognised by men of science. Even anyone with the slightest knowledge of biology, Professor Bateson remarked in a British Association Presidential address in 1914, is aware that a population need not be declining because it is not increasing; "in normal stable conditions population is stationary." Major Leonard Darwin, the thoughtful and cautious President of the Eugenics Education Society, has lately stated his considered belief ("Population and Civilisation," Economic Journal, June, 1921) that increase in numbers means, ultimately, relative reduction of wealth per head, with consequent lowering of the standard of civilisation; that it also, under existing conditions, involves the production of a smaller proportion of men of ability; and, further, a depreciation of our traditions; he concludes that, whatever element in civilisation we regard—wealth, or stock, or traditions—"any increase in the population such as that now taking place will be accompanied by a lowering in the standard of our civilisation."

There are definite reasons why real progress in the supreme tasks of civilisation can best be made by a more or less stationary population, whether the population is large or small, and it need scarcely be added that, so far as the history of mankind is yet legible, the great advances in civilisation have been made by small, even very small populations. Where the population is rapidly growing, even if it is growing under the favourable conditions that hardly ever accompany such growth, all its energy is absorbed in adjusting its perpetually shifting equilibrium. It cannot succeed in securing the right conditions of growth, because its growth is never ceasing to demand new conditions. The structure of its civilisation never rises above the foundations because these foundations have perpetually to be laid afresh, and there is never time to get further. It is a process, moreover, accompanied by unending friction and disorder, by strains and stresses of all kinds, which are fatal to any full, harmonious, and democratic civilisation. The "population question," with the endlessly mischievous readjustment it demands, must be eliminated before the great House of Life can be built up on a strong solid human foundation, to lift its soaring pinnacles towards the skies. That is what many bitter experiences are beginning to teach us. In the future we are likely to be much less concerned about "race-suicide," though we can never be too concerned about race-murder.

When we think, however, of the desirability of a more or less stationary population, in order to insure real social progress, as distinct from that vain struggle of meaningless movement to and fro which the history of the past reveals, we have to be clear in our minds that it may be far from desirable that the present overgrown population of the world should be stationary. That might indeed be better than further increase in numbers, it would arrest the growth of our present evils; it might open the way to methods by which they would be diminished or eliminated. But the process would be infinitely difficult, and almost infinitely slow, as we may easily realise when we consider that, with a population even smaller than at present, the human race has not only ravished the world's beauty almost out of existence, but so ravaged its own vital spirit that, as was found with some consternation during the Great War, a large proportion of the male population of every country is unfit for military service.

So often we hear it assumed, or even asserted, that greatness means quantity, so that to look forward to the replacement of the present teeming insignificant human myriads by a rarer and more truly greater race is to be a pessimist! Oh, these "optimists"! To revel in a world which more and more closely resembles all that the poets ever imagined of Hell, is to be an "optimist"! One wonders how it is that in no brief moment of lucidity it occurs to these people that the lower we descend in the scale of life the greater the quantity in a species and the poorer the quality, so that to reach what such people should really regard as the world's period of supreme greatness in life we must go back to the days, before animal life appeared, when the earth was merely a teeming mass of bacteria.[26]

[26] See, for instance, H.F. Osborn, The Origin and Evolution of Life, 1918, Chapter III.

To-day, we are often told, the majority of human beings belong either to the Undesired Class or the Undesirable Class. To realise that this is so, we are bidden to read the newspapers or to walk along the streets of the cities—whichever they may be—wherein dwell the highest products of our civilisation. In the better class quarters it is indeed the Undesirable Class that seems to predominate, and in the poor quarters, the Undesired. Yet, viewing our species as a whole, the two classes may be seen to walk hand in hand along the same road, and in proportion as our nobler instincts germinate and develop, we must doubtless admit that it ought to be our active aim to make that road for both of them—socially though not individually—the Road to Destruction.

To stem the devastating tide of human procreativeness, however, easy as it may seem in theory, is by no means so easy as some think, especially as those think who believe that the human race stands on the brink of suicide. For there is this about it that we must never forget: the majority of those born to-day die before their time, so that by diminishing the production of the unfit, as well as by the progressive improvement of the environment that automatically accompanies such diminution, we may make an imposing difference in the appearance of the birth-rate, whilst yet the population goes on increasing rapidly, probably even more rapidly than before. It needs a most radical and thorough attack on the birth-rate before we can make any real impression on the rate of increase of the population, to say nothing of its real reduction. There is still an arduous road before us.

True it is that we have two opposing schools of thought which both say that we need not, or that we cannot, make any difference by our efforts to regulate the earth's human population. According to one view the development of population, together with the necessity for war which is inextricably mixed up with a developing population, cannot be effected without, as one champion of the doctrine is pleased to put it, "shattering both the structure of Euclidean space and the psychological laws upon which the existence of self-consciousness and human society are conditional."[27] In simpler words, populations tend to become too large for their territories, so that war ensues, and birth-control can do nothing because "it is doubtful whether a group in the plenitude of vigour and self-consciousness can deliberately stop its own growth." The other school proclaims human impotence on exactly opposite grounds. There is not the slightest reason, it declares, to believe that birth-control has had any but a completely negligible influence on population. This is a natural process and fertility is automatically adjusted to the death-rate. Whenever a population reaches a certain stage of civilisation and nervous development its procreativeness, quite apart from any effort of the will, tends to diminish.

The seeming effect of birth-control is illusory. It is Nature, not human effort, which is at work.[28]

[27] B.A.G. Fuller, "The Mechanical Basis of War," Hibbert Journal, 1921.

[28] Sir Shirley Murphy some years ago (Lancet, 10 Aug. 1912) argued that the fall of the birth-rate, as also that of the death-rate, has been largely effected by natural causes, independent of man's action. Mr. G. Udney Yule (The Fall in the Birth-rate, 1920) also believes that birth-control counts for little, the chief factor being natural fluctuations, probably of economic nature. Recently Mr. C.E. Pell, in his book, The Law of Births and Deaths (1921), has made a more elaborate and systematic attempt to show that the rise and fall of the birth-rate has hitherto been independent of human effort.

These two opposing councils of despair, each proclaiming, though in a contrary sense, the vanity of human wishes in the matter of procreation, might well, some may think, be left to neutralise each other and evaporate in air. But it seems worth while to point out that, with proper limitations and qualifications, there is an element of truth in each of them, while, without such limitations and qualifications, both are alike obviously absurd and wrong-headed. Undoubtedly, as the one school holds, in certain stages of civilisation, even at a fairly advanced stage, nations tend to break out over their frontiers with resulting war; but the period when they reach "the plenitude of vigour and self-consciousness" is exactly the period when the birth-rate begins to decline, and the population, deliberately or instinctively, controls its own increase. That has, for instance, been the history of France since the great expansion of population, roughly associated with the Napoleonic epopee,—which doubtless covered a web of causes, sanitary, political, industrial, favourable to a real numerical increase of the nation— had died down slowly to the level we witness to-day.[29] Similarly, with regard to the opposing school, we must undoubtedly accept a natural fall in the birth-rate with a rising civilisation; that has always been visible in highly civilised individual couples, and it is an easily ascertainable zoological fact that throughout the evolution of life procreativeness has decreased with the increased development of species. We may agree that a natural factor comes into the recent fall in the human birth-rate. But to argue that because a natural decline in birth-rate is the essential factor in the slowing down of procreative activity with all higher evolution, therefore deliberate birth-control counts for nothing, since exactly the same result follows when voluntary prevention is adopted and when it is not, seems highly absurd. We must at least admit that voluntary birth-control is an important contributory cause, in some sense indeed, of supreme importance, because it is within man's own power and because man is thus enabled to guide and mould processes of Nature which might otherwise work disastrously. How disastrously is shown by the history of Europe, and in a notable degree

France, during the four or five centuries preceding the end of the eighteenth century when various new influences began to operate. During all these centuries there was undoubtedly a very high birth-rate, yet infant mortality, war, famine, insanitation, contagious diseases of many and virulent kinds, tended, as far as we can see, to keep the population almost or quite stationary,[30] and so ruinous a method of maintaining a stationary population necessarily used up most of the energy which might otherwise have been available for social progress, although the stationary population, even thus maintained, still placed France at the head of European civilisation. The more firmly we believe that the diminution of the population is a natural process, the more strenuously, surely, we ought to guide it, so that it shall work without friction, and, so far as possible, tend to eliminate the undesirable stocks of man and preserve the desirable. Clearly, the theory itself calls for much effort, since it is obvious that along natural lines the decline, if it is the result of high evolution, will affect the fit more easily than the unfit.

[29] The reader may point to the renewal of Militarism and Imperialism in France since the Great War. That, however, has been an artificial product (in so far as it exists among the people themselves) directly fostered from outside by the policy of England and the United States, just as the same spirit in Germany before the war, in the face of a falling birth-rate, was artificially fostered from above by a military and Imperialistic caste.

[30] See especially Mathorez, Histoire de la Formation de la Population Française, Vol. I, 1920, Les Étrangers en France. The fecundity of French families, even among the aristocracy, till towards the end of the eighteenth century, was fabulous; in the third quarter of the seventeenth century the average number of children was five in Paris. But the mortality was extremely high; under the age of sixteen, Mathorez estimates, it was 51 per cent., and infant mortality was terrible in all classes, small-pox being specially fatal. Then there were the various diseases termed plagues, with famine sometimes added, while war, emigration, and religious celibacy all counteracted the excessive fecundity, so that from the thirteenth century to the third quarter of the eighteenth the population seems to have been stationary, about twenty-two millions. Then the size of the family fell in Paris to 3.9 and in France generally to 4.3, while also there were fewer marriages. Therewith there was an increase of prosperity.

Thus there seems, on a wide survey of the matter, no reason whatever to quarrel with that conviction, which is gradually over-spreading all classes of human society in all parts of the world, and ever more widely leading to practical action, that the welfare of the individual, the family, the community, and the race is bound up with the purposive and deliberate practice of birth-control, whether we advocate that policy on the ground that we are thereby furthering Nature, or on the opposite, and no doubt equally excellent, ground that we are thereby correcting Nature.

Along this road, as along any other road, we shall not reach Utopia; and since the Utopia of every person who possesses one is unique that perhaps need not be regretted. We shall not even, within any measurable period of time, reach a sanely free and human life fit to satisfy quite moderate aspirations. The wise birth-controller will not (like the deliciously absurd suffragette of old-time) imagine that birth-control for all means a New Heaven and a New Earth, but will, rather, appreciate the delightful irony of the Biblical legend which represented a world with only four people in it, yet one of them a murderer. Still, it may be pointed out, that was a state of things much better than we can show now. The world would count itself happier if, during the Great War, only twenty-five per cent of the population of belligerent lands had been murderers, virtually or in fact. There is something to be gained, and that something is well worth while.

Still, whether we like it or not, the task of speeding up the decrease of the human population becomes increasingly urgent.[31] To many of our Undesirables it may seem, mere sentiment to trouble about the ravishing of the world's beauty or the ravaging of the world's humanity. But certain hard facts, even to-day, have to be faced. The process of mechanical invention continues every day on an ever increasing scale of magnitude. Now that process, however necessary, however beneficial, involves some of the chief evils of our present phase of what we call civilisation, partly because it has deteriorated the quality of all human products and partly because it has enslaved mankind, and in so doing deteriorated also his quality.[32] Now we cannot abolish machinery, because machinery lies in the very essence of life and we ourselves are machines. But, as the largest part of history shows, there is no need whatever for man to become the slave of machinery, or even for machinery to injure the quality of his own work; rightly used it may improve it. The greatest task before civilisation at present is to make machines what they ought to be, the slaves, instead of the masters of men; and if civilisation fails at the task, then without doubt it and its makers will go down to a common destruction. It is a task inextricably bound up with the task of moulding the human race for which birth-control is the elected instrument. Indeed they are but two aspects of the same task. We have to accept the rugged fact that every step to render more nearly perfect the mechanical side of life correspondingly abolishes the need for men. Thus it is calculated to-day that whenever, in accordance with a growing tendency, coal is superseded by oil in industry two men are enabled to do the work of twelve. That is merely typical of what is taking place generally in our modern system of civilisation. Everywhere a small number of men are being enabled to replace a large number of men. Not to avoid looking ahead, we may say that of every twelve millions of our population, ten millions will be unwanted. Let them do something else! we cheerfully exclaim. But what?

No doubt there are always art and science, infinite in their possibilities for joy and enlightenment, infinite also, as we know, in their possibilities of mischief and shallowness and boredom. Let it only be true science and great art, and one man is better than ten millions. To say that is only to echo unconsciously the ancient saying of Heraclitus, "One is ten thousand if he be the best."

[31] Professor E.M. East, a distinguished biologist and lately President of the American Society of Naturalists (Nature, 23 Sept., 1920), has estimated that, for all the fall in the birth-rate, the present rate of increase in the population of the world, chiefly of whites, who are increasing most rapidly, will, in the lives of our grandchildren, lead to a struggle for existence more terrible than imagination can conceive.

[32] This has been set forth with admirable lucidity and wealth of illustration by Dr. Austin Freeman in his Social Decay and Regeneration (1921), already mentioned.

The vistas that are opened up when we realise the direction in which the human race is travelling may seem to be endless; and so in a sense they are. Man has replaced the gods he once dreamed of; he has found that he is himself a god, who, however realistic he seeks to make his philosophy, himself created the world as he sees it and now has even acquired the power of creating himself, or, rather, of re-creating himself. For he recognises that, at present, he is rather a poor sort of god, so much an inferior god that he is hardly, if at all, to be distinguished from the Lords of Hell.

The divine creative task of man extends into the future far beyond the present, and we cannot too often meditate on the words of the wisest and noblest forerunner of that future: "The whole world still lies before us like a quarry before the master-builder, who is only then worthy of the name when out of this casual mass of natural material he has embodied with all his best economy, adaptability to the end, and firmness, the image which has arisen in his mind. Everything outside us is only the means for this constructing process, yes, I would even dare to say, also everything inside us; deep within lies the creative force which is able to form what it will, and gives us no rest until, without us or within us, in one or the other way, we have finally given it representation." The future, with all its possibilities, is still a future infinitely far away, however well it may be to fix our eyes on the constellation towards which our solar system may seem to be moving across the sky.

Meanwhile, every well-directed step, while it brings us but ever so little nearer to the far goal around which our dreams may play, is at once a beautiful process and an invigorating effort, and thereby becomes in itself a desirable end. It is the little things of life which give us most satisfaction and the smallest things in our path that may seem most worth while.

www.ingramcontent.com/pod-product-compliance
Lightning Source LLC
Chambersburg PA
CBHW072249310526
45795CB00011B/575